Yiming Sun

Adaptive and Intelligent Temperature Control of Microwave Heating Systems with Multiple Sources

Karlsruher Forschungsberichte aus dem
Institut für Hochleistungsimpuls- und Mikrowellentechnik

Herausgeber: Prof. Dr.-Ing. John Jelonnek

Band 8

Adaptive and Intelligent Temperature Control of Microwave Heating Systems with Multiple Sources

by
Yiming Sun

KIT Scientific Publishing

Dissertation, Karlsruher Institut für Technologie (KIT)
Fakultät für Elektrotechnik und Informationstechnik, 2015
Referenten: Prof. Dr. rer. nat. Marc Weber
Prof. Dr.-Ing. John Jelonnek

Impressum

KIT Scientific Publishing

Karlsruher Institut für Technologie (KIT)
KIT Scientific Publishing
Straße am Forum 2
D-76131 Karlsruhe

KIT Scientific Publishing is a registered trademark of Karlsruhe Institute of Technology. Reprint using the book cover is not allowed.

www.ksp.kit.edu

Print on Demand 2016

ISSN 2192-2764
ISBN 978-3-7315-0467-2
DOI: 10.5445/KSP/1000051503

Foreword of the Editor

Microwave heating is becoming popular in industrial processes. Different to conventional heating techniques (e. g. hot air, infrared light), the microwave heating principle is volumetric, which means the microwave penetrates through the object and is heating its volume uniformly. Hence, both surface and interior of the object are heated simultaneously, which results in significantly reduced process time and energy consumption.

However, the microwave heating is not always uniform. Unlike in convectional heating where the temperature distribution is reaching equilibrium after a certain process time, the temperature distribution of microwave heating is varying during the entire heating process. There are a number of spatially distributed hot spots and cold spots, caused by not only the properties of the workpiece, but also the electromagnetic (EM) field distribution within the oven chamber.

In his work, Dr.-Ing. Yiming Sun proposes an innovative real-time microwave control approach to improve the temperature homogeneity under microwave heating by combining the idea of advanced control algorithms and multi-source microwave feeding systems. In this approach, Dr.-Ing Yiming Sun develops multiple adaptive or intelligent control structures, including the model predictive control (MPC), neural network control (NNC) and reinforcement learning control (RLC) methods, to automatically adjust the power of spatially distributed feeding antennas, in order to actively influence the electromagnetic (EM) field distribution inside the oven chamber, and hence, improve the temperature distribution. The experimental results prove that these advanced control methods can effectively reduce the final temperature derivations and improve the temperature homogeneity, compared with conventional feedback control strategies.

The work of Dr.-Ing. Yiming Sun is showing a big potential to be implemented in industry applications, especially for those requiring excellent temperature homogeneity during the heating process. It will help the multi-source microwave heating devices to achieve the same performance as common industry autoclaves, in terms of temperature homogeneity. It will also further promote the implementation of microwave heating.

Adaptive and Intelligent Temperature Control of Microwave Heating Systems with Multiple Sources

Zur Erlangung des akademischen Grades eines

DOKTOR-INGENIEURS

von der Fakultät für
Elektrotechnik und Informationstechnik
des Karlsruher Institut für Technologie (KIT)

genehmigte

DISSERTATION

von

Yiming Sun

geb. in Shandong, China

Tag der mündlichen Prüfung: 08.06.2015

Hauptreferent: Prof. Dr. rer. nat. Marc Weber
Korreferent: Prof. Dr. -Ing. John Jelonnek

Contents

List of Symbols

∇	Gradient operator
∂	Partial derivative operator
d	Ordinary derivative operator
$\|\|\ \|\|$	Euclidean norm of vectors
$\mathbf{A}^{\mathrm{T}}$	Transpose of vector/matrix
$\vec{\mathbf{e}}_{\mathrm{x/y/z}}$	Unit vector of a Cartesian coordinate system in the x/y/z-direction

In the electromagnetic heating part (chapter 2), scalars are represented by normal symbols and vectors are represented by capital bold symbols with arrows.

$\vec{\mathbf{E}}$	Electric field	$\mathrm{V/m}$
$\vec{\mathbf{D}}$	Electric displacement	$\mathrm{C/m^2}$
$\vec{\mathbf{H}}$	Magnetic field strength	$\mathrm{A/m}$
$\vec{\mathbf{B}}$	Magnetic flux density	T
$\vec{\mathbf{J}}_{\mathrm{r}}$	Free current density	$\mathrm{A/m^2}$
ρ_{f}	Free charge density	$\mathrm{C/m^3}$
ω	Angular frequency	rad/s
f	Operating frequency	Hz
f_n	Resonant frequency for the mode n in a microwave applicator	Hz
Q_n	Quality factor for the mode n in a microwave applicator	unitless

Q_{L}	Quality factor corresponding to the energy loss caused by the load	unitless
Q_{W}	Quality factor corresponding to the energy loss caused by the wall	unitless
Q_{D}	Quality factor corresponding to the energy loss caused by diffraction	unitless
τ	Relaxation time of the dielectric material	s
γ	Complex propagation vector	1/m
α	Attenuation constant (real part of γ)	1/m
β	Phase constant (imaginary part of γ)	1/m
ε	Complex permittivity of material	F/m
ε_{s}	Approximation of the complex permittivity at low frequency range	F/m
ε_{∞}	Approximation of the complex permittivity at very high frequency range	F/m
ε'	Real part of the complex permittivity	F/m
ε''	Imaginary part of the complex permittivity	F/m
ε_0	Permittivity of vacuum	F/m
ε_{r}	Relative permittivity of material	F/m
μ_0	Permeability of free space	H/m
μ_{r}	Relative permeability of material	H/m
T	Temperature of the load	K
ρ	Mass density	kg/m^3
c_{p}	Specific heat capacity	J/ (K · kg)
κ	Thermal conductivity	W/ (m · K)
$\sigma_{\mathrm{e}}(T)$	Electrical conductivity at temperature T	A/ (V · m)

σ'	Stefan-Boltzmann constant	$\mathrm{W}/\left(\mathrm{m}^2 \cdot \mathrm{K}^4\right)$
h	Convection heat transfer coefficient	$\mathrm{W}/\left(\mathrm{m}^2 \cdot \mathrm{K}\right)$
ϱ	Emissivity of the surface of the heated load	unitless

In the mathematical modeling and control parts (chapter 3 and 4), scalars are represented by non-bold symbols. Vectors are represented by bold symbols, and matrices are represented by bold symbols with brackets.

l	Length of unit cell	m
p_{cd}	Dissipated/heating power density by conduction	$\mathrm{W/m^3}$
p_{cv}	Dissipated power density by convection	$\mathrm{W/m^2}$
p_{rd}	Dissipated power density by radiation	$\mathrm{W/m^2}$
p_{mw}	Microwave heating power density	$\mathrm{W/m^3}$
P_{cd}	Dissipated/heating power by conduction	W
P_{cv}	Dissipated power by convection	W
P_{rd}	Dissipated power by radiation	W
P_{mw}	Microwave heating power	W
T_{a}	Ambient temperature	K
Y	Temperature difference between the load and ambient air	K
$\mathbf{T}$	Temperature vector of the heated load	K
$\mathbf{Y}$	Temperature difference vector	K
$\mathbf{U}$	Power portion vector	unitless
$\mathbf{V}$	Amplitude portion vector	unitless
$\mathbf{P}$	Dissipated/heating power vector	unitless

$\mathrm{E}^{n}_{\mathrm{x/y/z}}$	Components of $\overrightarrow{\mathbf{E}}_n$ in the x/y/z-direction	V/m
$\mathrm{E}^{n,m}_{\mathrm{x/y/z}}$	Components of microwave from the m-th feeding source to the n-th location in the x/y/z-direction	V/m
$\mathbf{E}^{n}_{\mathrm{x/y/z}}$	Vector of components from all sources to the n-th location in the x/y/z-direction	V/m

In continuous-time model:

$A_{\mathrm{c}}(t), A^{n}_{\mathrm{c}}(t)$	Self-cooling factor (of the n-th temperature)	1/s
$[\mathbf{A}_{\mathrm{c}}(t)]$	Self-cooling matrix	1/s
$B^{n,m}_{\mathrm{c}}(t)$	Temperature changing rate of the n-th location caused by the m-th feeding source	K/s
$\mathbf{B}^{n}_{\mathrm{c}}(t)$	Temperature changing rate vector including all $B^{n,m}_{\mathrm{c}}(t)$ with $1 \leq m \leq M$	K/s
$[\mathbf{B}_{\mathrm{c}}(t)]$	Temperature changing rate matrix including all $B^{n,m}_{\mathrm{c}}(t)$ with $1 \leq m \leq M$ and $1 \leq n \leq N$	K/s
$[\mathbf{\Phi}^{n}_{\mathrm{c}}(t)]$	Effective heating rate matrix at the n-th location	K/s

In discrete-time model:

$A(k), A^{n}(k)$	Self-cooling coefficient (of the n-th temperature)	unitless
$[\mathbf{A}(k)]$	Self-cooling matrix in a discrete-time form	unitless
$B^{n,m}(k)$	Temperature changing of the n-th location caused by the m-th feeding source	K
$\mathbf{B}^{n}(k)$	Temperature changing vector including all $B^{n,m}(k)$ with $1 \leq m \leq M$	K

$[\mathbf{B}(k)]$	Temperature changing matrix including all $B^{n,m}(k)$ with $1 \leq m \leq M$ and $1 \leq n \leq N$	K
$\mathbf{\Phi}^n(k)$	Temperature increasing matrix at the n-th location	K
$\mathbf{\Psi}(k)$	Temperature increasing vector of MIMO form	K

In controller design:

$J(k)$	Cost/objective function to be optimized at time k
$[\mathbf{\Gamma}]$	Positive definite tuning matrix used in the cost function
$\mathbf{V}_i^{\mathrm{c}}$	The i-th initialized control sequence in the GA algorithm
$\mathbf{Y}_i^{\mathrm{c}}$	Predicted temperatures using $\mathbf{V}_i^{\mathrm{c}}$
$\mathbf{w}$	Weights vector/matrix used in the NN
$S(k)$	The state of the MDP at time k
$R(k)$	The reward received from the plant at time k
$\pi(s, \mathbf{u})$	Control policy used in RL
$V_\pi(s)$	The state value function under the state s using the policy π
$Q_\pi(s, \mathbf{u})$	The state-action value function of the state-action pair $(s, \mathbf{u})$ using the policy π
Δ_{td}	The temporary difference error
$e(s, \mathbf{u})$	The eligibility trace defined in RL

List of Figures

List of Tables

1. Introduction

Heating is one of the most common processes in the manufacturing industry for a wide range of applications, including sintering, curing and drying. It plays an essential role in promoting chemical and physical reactions, and improving the processability of materials. The quality of heating directly determines the final quality of industrial products. In conventional industrial heating equipments, such as furnaces, kilns and autoclaves, heat is generated by the combustion of fuels or using electric currents, and transferred to the heated material by either thermal conduction, convection, radiation, or a combination of all three.

In conventional heating, all the heating energy required will be firstly generated on the surface by the thermal radiation or convection, and then transmitted through the body into the core by the thermal conduction. The rate of the heat flow from outside to inside is entirely determined by the physical properties of the material, such as its conductivity, specific heat capacity and mass density. There is no way to accelerate the heating flow rate or influence the internal temperature distribution. With the same material, the thicker the heated product is, the longer it takes to transfer the heat to the internal core. Meanwhile, a large temperature deviation would occur between the surface and internal core, which could bring quality problems [Mer98]. During the whole heating process, a huge amount of energy is lost in heating the intermediate medium (e.g. hot air), and the corresponding time consumption is also large. Although many proper improvements have been done regarding to equipment sizes and operations, conventional heating has various problems unsolved. Heating still remains as one of the most challenging processes to be precisely controlled.

1.1. Development of Microwave Heating

In contrast to conventional heating, microwave heating has a completely different heating mechanism, which is the so-called volumetric heating ability. Instead of the heating flow from the surface to the core, the microwave energy is directly transferred into heat inside the heated material.

As shown in the figure 1.1, in conventional heating, the surface is always hotter than the core because the heat flows from the surface to the core. However, in microwave heating both the surface and the core of the product are heated simultaneously. Due to the heat convection on the surface, the maximum temperature of microwave heating occurs in the core of the product. It is the opposite to the temperature distribution of the conventional heating, therefore it is called the inverse temperature profile. This inverse temperature profile might be a problem for some applications, but it is helpful in certain cases, such as in the microwave sintering of YBCO superconductors [BADAC92].

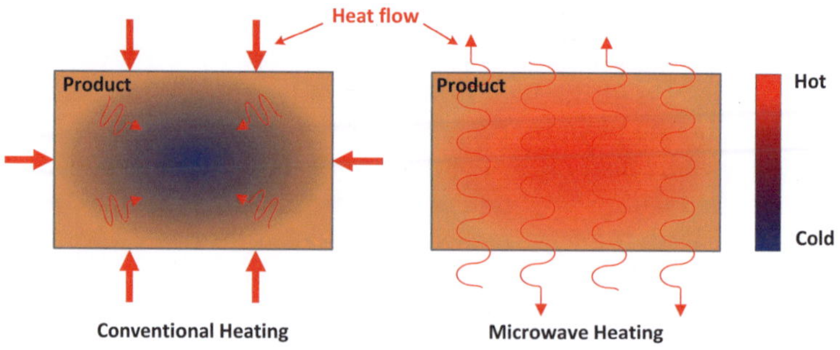

Figure 1.1. Temperature profile comparison between conventional (left) and microwave (right) heating.

Benefiting from the special heating mechanism, microwave heating has several advantages over the conventional heating.

1. Reduced energy and time consumption

   Microwave heating generates the heat directly in the volume of a product, which saves significant energies from the unnecessary heating of the intermediate medium, especially for high temperature applications [TC99]. Because of the unique volumetric heating ability, there is no need to spend additional time in the surface heating and the heat transfer process. Moreover, the heating rate to the inside of the material is not limited by the slow heat transfer process, but mostly determined by the microwave power. In principle, the microwave heating is able to provide any desired heating rate as long as the microwave power is sufficient, which largely reduces the overall heating time [LM03].

2. Selective and controllable heating

   When a product consisting of different materials is heated in a microwave cavity, different heating rates are achieved due to different properties of individual materials. This selective heating phenomenon is desired in many applications, especially in the microwave chemistry area. For example, in the applications introduced by [ZQY14], temperature gradients at different locations are required to enhance different local chemical reactions.

   In conventional heating, it is slow and clumsy to control the heating rate. In comparison to the conventional heating, the rate of microwave heating can be immediately controlled by adjusting the microwave power. In applications where multiple stages with different heating rates are required, i.e. sintering of ceramics [Kat92] or curing of composite materials [LLG14], microwave heating is preferred because of its controllable and quick response.

Due to above mentioned advantages, considerable research has been performed on microwave heating since the 1950s. Nowadays, there are a large number of successful applications of microwave heating in different industries, including food, rubber, pharmaceutical, polymers, plastics and textiles [oED07]. During the last 10 to 15 years, more and more interests have been drawn to the emerging domain of fabricating carbon fiber reinforced polymer (CFRP) composite materials using microwave heating. CFRP composite material is one kind of advanced high-performance (AHP) composite materials that increasingly attracts both commercial and research attentions [Cor01]. Due to its excellent mechanical properties, such as the fatigue resistance, light weight and extraordinary high-temperature characteristics [LLG14], CFRP has been widely applied in automotive, renewable energy, aerospace and many other weight-critical applications. According to the report in [Qui01], nearly half of the Boeing 787 airplane is comprised of CFRP and other composites, which saves on average 20% of the total weight compared with conventional aluminium designs. Based on the market report published in 2012 [Cas12], the global CFRP market size is forecast to have an annual growth rate of 17%, reaching 118,600 tonnes with an estimated value of 7.3 billion US dollars (USD) by 2017.

There is no doubt that CFRP will become one of the most important and demanding materials in the near future. However, there are also challenges from different aspects, such as high costs and manufacturing complexities. For most industrial areas, cost reduction is a key priority that has to be concerned during the manufacture process. Although the price of carbon fiber material has been decreasing during the last few years (about 20 USD per kg in 2014 [cfr]), curing of one CFRP workpiece is still expensive. The most common way to cure a CFRP workpiece is to layer the uncured prepreg into a mould, put a vacuum bag to the mould and cure it in an autoclave. The curing process within the autoclave takes a long time and the corresponding energy consumption is large, which leads to the increase of the total cost. Besides the autoclave curing, there are a number of other curing methods. Their main advantages and drawbacks are shown in the following table 1.1.

Compared with other curing methods in table 1.1, microwave heating has a higher heating efficiency and lower curing cost. In addition, microwave heating also has a large penetration depth, which means the heat can be directly generated deeply inside the CFRP workpiece and rapid curing is possible for thick composite materials. However, there are still a number of problems that limit the implementation of microwave heating in the practical CFRP curing process, such as thermal insulation, temperature measurement, and most important, temperature control problems.

During the heating process, the temperature of the entire heated workpiece should be controlled and held in a predefined range. In addition, the overall temperature homogeneity of the workpiece should also be sustained in a certain level, to avoid any local overheating (hot spots) [RMS96] or thermal runaway [Kri92]. According to investigations in

	Advantages	Drawbacks
Autoclave	high quality workpiece	expensive; slow
Electron-beam	high quality workpiece; low costs; short time	very large temperature difference for thick workpieces
Ultra violet (UV)	very short time	special resin matrix needed; relatively high costs
X-rays	large penetration depth	long curing time; safety issues
Microwave Heating	short time; low costs; large penetration depth; high heating efficiency	temperature control and non-uniform distribution problems

Table 1.1. Comparison of different curing methods (according to [BRS01]).

[LLG14] and [KMS01], quality failures caused by the non-uniform temperature distribution have become the major problem associated with microwave heating applications.

It is easy to control a single temperature under microwave heating, by using a temperature sensor and adjusting the microwave power based on the feedback temperature value [SZ06]. Nevertheless, it is way more complicated and difficult to control the entire temperature distribution of the workpiece. Because the temperature distribution is determined not only by the properties of the workpiece, but also by the electromagnetic (EM) field distribution within the microwave cavity. The EM field and temperature distributions are governed jointly by two partial differential equations (PDEs) [BBBB95], which are the Maxwell's equations [Mer98] and the forced heat transfer equation [Met96]. On the one hand, the EM field distribution influences the heating rates as well as the temperature distribution of the workpiece. On the other hand, the temperature distribution also affects the properties of the workpiece, such as thermal conductivity and dielectric permittivity, which is reciprocated to change the EM field distribution. An uniform temperature distribution of microwave heating requires optimizations from a number of different parts, including the design of the microwave applicator, adjustment of microwave power, temperature measurement approach and others.

1.2. Overview of the State of the Art

In general, there are three different types of approaches to control the temperature and improve the temperature homogeneity under microwave heating.

1.2.1. Modern control methods

The idea of using modern control methods to control the heat transfer process in microwave heating started from the 1980s. In the beginning it was used simply as an approach to keep the thermal stability and prevent the thermal runaway phenomenon. In simple scenarios, e.g. one-dimensional slabs or single-mode microwave cavities, the two coupling PDEs (Maxwell's equations and heat transfer equation) can be solved analytically, and the required microwave power is directly calculated based on the analytical solution, such as the idea used in [Smy90]. But this control scheme is only valid for simple heating scenarios, and it also assumes properties of the heated material are perfectly known in advance, which does not hold in most cases.

Besides the analytical control solution method, another more general control approach, which is the model based temperature control, was proposed in [RMTV85]. A transfer function [Dor95] was used to describe the relationship between one measured temperature and the microwave power of one source. A simple single-input single-output (SISO) feedback controller was constructed to control the power of the microwave source in order to make the measured temperature reach the predefined target temperature.

This modeling-and-control approach is effective for microwave heating systems with one feeding source, and most of following work uses the same kind of strategy. For instance, in [RCVI99] a proportional-integral-derivative (PID) controller was developed to control temperatures in microwave heating. In [SBA00] the autoregressive-moving-average model with exogeneous inputs (ARMAX) [BJR13] method was utilized to build the system model. Another model-based feedback controller was designed and a recursive Kalman filter (RKF) was used to estimate the parameter of the ARMAX model, which further enhanced the control performance. Similar control schemes can also be found in [KRS08] and [HS07]. Despite the model based feedback controller achieved good control performance regarding to the SISO heating scenario, the entire temperature distribution of the heated load was not considered in the control model, and hence, not controlled.

In order to truly improve the temperature homogeneity, microwave heating systems equipped with multiple distributed feeding sources were studied and developed, such as the system used in our experiments and the one mentioned in [LLHG14]. On the one hand, systems with multiple distributed microwave heating sources have a great potential to achieve a more uniform temperature distribution. But on the other hand, they are much more complicated than systems with only one feeding source. A lot of its properties, such as the EM distribution, excitement of different modes, effects of different positions of sources, especially the temperature control strategy, still remain unknown. Although in [LLG14] and [LLHG14], a distributed temperature control system was claimed to be implemented for controlling power of individual sources, there is no detailed introduction about this system. Corresponding modeling and controlling methods are still not clear. To the author's knowledge, up until now there is no sophisticated theory or model published to deal with the controller design of systems with multiple distributed microwave heating systems.

1.2.2. Numerical simulations

Numerical simulation is preferred for more complicated scenarios where analytical solutions are not feasible. The idea of numerical simulation is to separate the original problem into a number of small parts, solve different parts individually and finally merge results of all small parts into the solution of the original problem. During the last 20 years, the rapid development of powerful computers has led to more and more implementations of numerical analysis for solving complex electrodynamics problems. Common numerical simulation methods used for modeling microwave heating scenarios consist of the finite-difference time-domain (FDTD) method, the finite-volume time-domain (FVTD) method, the finite element method (FEM), the method of moments (MoM), as well as the transmission line matrix (TLM) method. Detailed introductions of above methods exceed the scope of this dissertation, and they can be found in a large number of books and papers such as [KL93] [LNS04] [DLT12] [Chr05].

Numerical simulation is used not to directly control the microwave heating, but to help improve the heating homogeneity. The principle is to firstly use numerical techniques to simulate and estimate the EM field as well as the temperature distributions within the microwave cavity, and then to optimize the heating homogeneity according to the simulation results using some auxiliary approaches. For example, in [CZ05] and [SCV+10], FDTD methods were used to simulate temperature profiles of microwave heating in one-dimensional and three-dimensional scenarios, and the thermal runaway phenomenon can be prevented based on the simulation. In [GRD07] FEM was applied to estimate the temperature distribution of the food within the microwave oven in case of a rotating turntable. Afterwards the rotating speed and time of the turntable can be optimized based on the simulation results. Hence, the overall temperature uniformity was enhanced. Similarly, the influences of mode stirrers to the EM field distribution were simulated using FEM in [PGMCCCSH04]. The angle of mode stirrers was determined with the help of simulation results and the temperature distribution was also improved. More examples of using numerical simulations to improve the performance of microwave heating can be found in [PS91] [FR93] [YG04] [GY07] [HM96].

Numerical simulation is the most important tool to analyze the EM field and temperature distributions of microwave heating, which is flexible and powerful to deal with different kinds of microwave heating problems. It provides important and valuable foundations to optimize the design and setup of the microwave heating system. Nevertheless, it is still not a direct and efficient way to improve the temperature homogeneity. On the one hand, the real EM and temperature distributions are affected by multiple factors, including dielectric properties of the heated product, the position of the heated product, the resonant microwave frequency range, and the air flow rate within the microwave cavity. Any little mismatch of these influencing factors will cause that the simulation result dramatically differs from the real heating result. Therefore, the more variables a problem has, the less accurate the simulation result is. That is also why in many problems with highly complex heating scenarios, the simulated results are far from satisfactory.

On the other hand, for certain highly complex systems (such as the HEPHAISTOS system used in this dissertation), even though the simulation results are accurate, there is no sophisticated guideline of how to improve the final temperature distributions based on corresponding simulation results. For instance, if a local hot spot is found in both the simulated and real temperature distributions, great efforts have to be done to eliminate this hot spot without creating new hot spots, due to the fact that any small modifications of the heating setup or equipment are possible to cause unexpected impacts to the entire temperature distribution.

1.2.3. Other auxiliary approaches

Besides the modern control method and the numerical simulation approach, there are other auxiliary approaches that aim to improve the temperature homogeneity under microwave heating. They can be divided into two categories. The first category comprises methods that focusing on creating a more homogeneous EM field distribution, which correspondingly leads to a homogeneous temperature distribution (for homogeneous workpieces). For example, the idea of variable frequency microwave heating was used in both [LBC+95] and

[Bow99], and the phase controlled microwave heating method was applied in [BPJD99]. According to the results shown in the three papers, although relatively homogeneous EM field distributions can be created using above methods, the resulting temperature distributions were not as uniform as expected, and local hot spots still existed.

The other category relies on mechanical approaches that can improve the temperature homogeneity directly. For instance, the method in [WRK+12] used the real experimental results to adjust the position of a susceptor, in order to achieve a specified heating at particular areas. Another example is the hybrid system used in [KHYM13] that consists of both the microwave and hot air heatings, to get a enhanced surface heating and reduce the temperature gradient between the surface and core of the product. These approaches are intuitive and effective. But on the other hand, they are either too difficult or not efficient enough to be extended to more general cases.

Despite numerous approaches were applied, no systematical solution has been established to solve the problem of the temperature non-uniformity in microwave heating. So far it still remains as one of the most challenging problems in microwave heating.

1.3. Objective and content

In this dissertation, a novel adaptive temperature control concept for microwave heating is proposed, which combines the distributed microwave feeding system with advanced multiple-input multiple-output (MIMO) control methods. The idea of distributed microwave feeding systems was presented and proved to be effective in [Feh09]. Based on this theory, HEPHAISTOS, which is short for high electromagnetic power heated automated injected structure oven system [Feh09], was designed. HEPHAISTOS has more than one microwave feeding sources and a very unique hexagonal structured cavity shape, resulting in a rather homogeneous EM field distribution in the whole cavity scale [Feh09]. More important, the EM field distribution in HEPHAISTOS can be changed during the heating process, by adjusting the power of individual heating sources, to achieve better temperature distributions for different heating setups or workpieces. In other words, HEPHAISTOS has the potential for systematically altering heating patterns to achieve the desired heating rates and temperature distributions on different types of heated workpieces.

The key to realize this potentiality is to develop an innovative and effective power control system. By means of advanced modeling and control methods, the temperature changing trends of different parts of the workpiece can be predicted and the power allocated to each heating source can be adjusted accordingly, to guarantee the heating stability and enhance the temperature homogeneity.

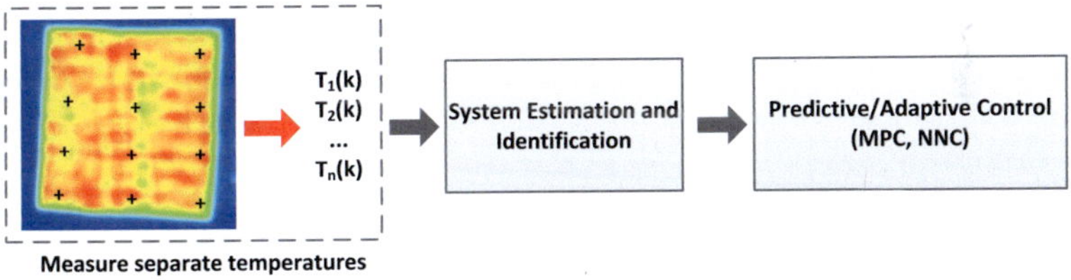

(a) Adaptive temperature control

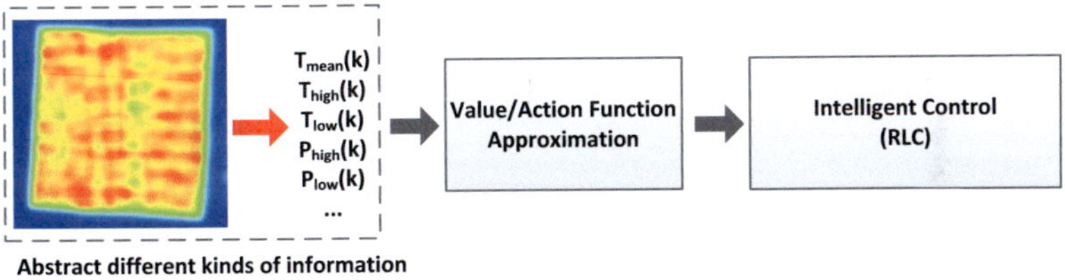

(b) Intelligent temperature control

Figure 1.2. Diagrams of two different temperature control approaches.

Due to the demands of different heating scenarios, two different control strategies are developed in this dissertation (see figure 1.2). Both of these two strategies aim to control the heating rate and achieve a desired temperature distribution, but they control based on different types of models. In the first adaptive temperature control approach (figure 1.2a), the modeling is based on temperatures measured at separate points. During the heating process a system model based on the temperature-power relation is recursively estimated, and the power of the distributed feeding sources is calculated based on the estimated model. The entire temperature distribution is effectively improved by driving separate temperatures to the predefined target temperature. This is the most common control strategy used in industrial applications. It is suitable for heating scenarios like the heating of multiple small workpieces or situations where the whole temperature profile of the workpiece is not available. Different system modeling methods, including the state-space model [Dor95] and the neural network (NN) based models [Hay98], are constructed. The model predictive control (MPC) [MHL99] and neural network control (NNC) [CK92] methods are applied, respectively.

The second intelligent temperature control approach (figure 1.2b) is based on the thermal image from the infrared camera. Information abstracted from the thermal image, such as the maximum temperature, the minimum temperature and their corresponding positions, is used for the control. Unlike the conventional input-output models used in the adaptive control system, a Markov decision process (MDP) [Bel57] is constructed to simulate the power-temperature state transient relation in the intelligent control system. The final control objective is to limit the temperatures of the entire workpiece within a predefined range, and hence, obtain a homogeneous temperature distribution. Conventional control methods are not appropriate for this task, therefore a reinforcement learning based intelligent controller [Bar98] is used here.

Comparing above two temperature control approaches with temperature controllers proposed in other papers such as [SAB98] [SBA00] [RCVI99] [HS07], the novel adaptive temperature concept developed in this dissertation has evident advantages in many aspects. Benefiting from the MIMO heating and control systems, it is possible to not only control the heating rate but also improve the temperature homogeneity in real time. The online system identification methods used in this control concept able to estimate influences from different factors numerically, without knowing any specific value of the influencing factor, which makes them more flexible be implemented in practical heating applications.

Last but no least, the theory and principles used in this dissertation are adaptive to other microwave heating systems and scenarios. The entire control system of HEPHAISTOS can be transferred to other microwave heating systems without significant modifications. All aforementioned methods and strategies will be introduced in details, and their corresponding performance will be compared via a number of experimental results.

2. Introduction of HEPHAISTOS

In this chapter, the distributed microwave heating system HEPHAISTOS will be introduced in detail. In the beginning, the principles of electromagnetic heating is addressed. Then the components of HEPHAISTOS, including the microwave heating system, the process control system and the temperature measurement system, are explained one by one.

2.1. Electromagnetic Heating

Almost everything in the real world, except a perfect conductor or perfect insulator, can be heated within an electromagnetic field. Based on different kinds of materials, the electrical heating techniques can be divided into several different classes, including ohmic heating, induction heating and dielectric heating.

2.1.1. Ohmic Heating

Ohmic heating, also known as Joule heating [GC99], is a basic electrical heating mechanism. When a direct current (DC) or alternating current (AC) flow passes through a conductor or semiconductor, a fraction of the electrical energy is transformed into thermal energy due to electrical resistance, and therefore the heat is generated. The amount of heating power P_{ohmic} is proportional to the square of the current I, such as

$$P_{\text{ohmic}} = I \cdot V = I^2 \cdot R = \frac{V^2}{R}, \tag{2.1}$$

where R is the electrical resistance of the object and V is the voltage across it. Ohmic heating is widely implemented in various industrial applications, including food processing [DAF90] and equipment manufacture (electric stove, electric heater) [WCJ$^+$12].

2.1.2. Induction heating

Induction heating is used for heating conducting materials (mostly metals). A conventional induction heating system is shown in figure 2.1. An alternating current passes through the induction coil and generates a time-variant magnetic field surrounding the coil. The magnetic field has the same frequency as the alternating current and it will induce eddy currents [RLCB02] within the conducting material that is inserted in the coil. Heat is generated by the eddy currents due to the ohmic heating effect (such as in equation 2.1). The frequency of AC power supply of induction heating is selected depending on the size of the workpiece to be heated. Heating smaller workpieces requires higher operating frequencies. Larger parts can be processed with lower operating frequencies. Normally the frequency ranges from 5 kHz to 500 kHz. Details about induction heating can be found in [Mer98] [RLCB02] and [RP06].

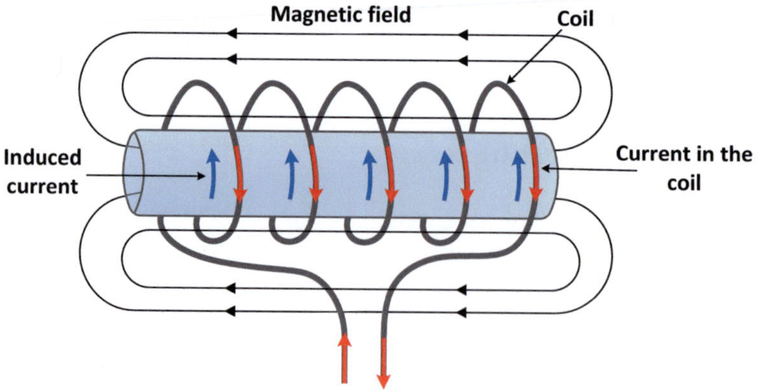

Figure 2.1. Induction heating system [RLCB02].

2.1.3. Dielectric heating

Dielectric heating is used to heat poorly conducting materials such as dielectrics and insulators, i.e. plastic, wood and rubber. According to the electrical frequencies used, dielectric heating is further classified into radio frequency heating and microwave heating. The frequency used in radio frequency heating is 300 kHz to 300 MHz, and in microwave heating is 300 MHz to 300 GHz [Met96]. The reason why dielectric heating is distinguished by the applied frequency is because at radio frequency the ionic conduction mechanism dominates the loss, whereas at microwave frequency the dipole relaxation is more important [Met96] [MM83]. In industry, the operating frequency of microwave heating is defined by ISM (the industrial, scientific and medical) frequencies as 915 MHz or 2.45 GHz (in Europe). Higher frequencies like 24.15 GHz could also be used, but it has to be verified with respect to special practical or theoretical advantages [Feh09].

When a dielectric material is put into an alternating electric field, the movements of permanent dipoles and free ions or ionic species are both affected, resulting in two different loss mechanisms: the dipole relaxation and the ionic conduction. In an alternating electric field, dipoles will rotate their orientations around equilibrium status to follow the external electric field, which is called dielectric polarization [GC99] such as illustrated in figure 2.2. Due to the time needed for the rotation, the response of re-orientation following external electric field is not instantaneous. At low frequencies, the dipoles have sufficient time to follow the direction altering of the electric field. There is little heat generated due to frictions during the rotation process, and most of the energy from the external electric field is directly stored in the dielectric.

As the frequency increases, the time left for the rotation gets less and less and finally it is shorter than the time needed for the rotation. It causes the dipole re-orientation lags the external electric field and this delay is called dipole relaxation [HS92]. In this case, the dipole alignment within the dielectric is broken and molecules collides with each other more and more. Correspondingly, frictions get larger and more heat is generated during the re-orientation process. As the frequency further increases, a critical point is reached where the rate of direction altering of the external electric field is so high that the dipole re-orientation fails to follow. As a consequence the friction becomes smaller and the generated heat is also less.

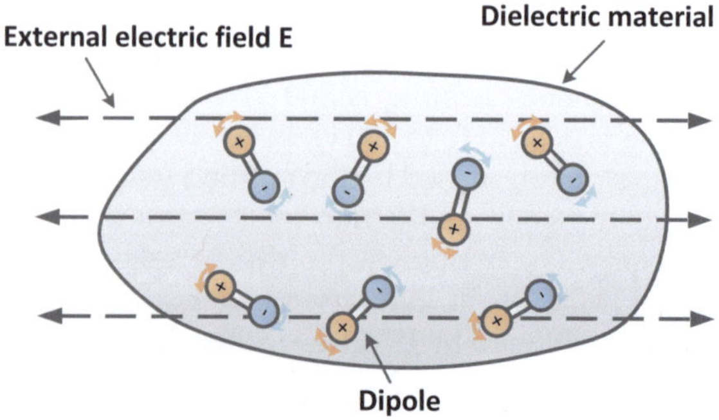

Figure 2.2. Dipole polarization of dielectric.

The above interpretation was firstly formulated by the physicist Peter Debye [Aru10], using the well-known complex permittivity equation [GC99], such as

$$\varepsilon(\omega) = \varepsilon'(\omega) - \mathrm{j}\varepsilon''(\omega) = \varepsilon_\infty + \frac{\varepsilon_s - \varepsilon_\infty}{1 + \mathrm{j}\omega\tau}, \tag{2.2}$$

where $\varepsilon(\omega)$ is the complex permittivity of a dielectric as a function of angular frequency $\omega = 2\pi f$, $\varepsilon'(\omega)$ and $\varepsilon''(\omega)$ are the real and imaginary part of $\varepsilon(\omega)$ respectively. The parameter ε_s is the permittivity of the material at low frequencies, ε_∞ is the permittivity of the material at very high frequencies. The relaxation time τ represents the time of form and decay of the polarization within the dielectric when the external electric field is applied and vanished respectively.

From the above equation, it can be derived that

$$\begin{aligned} \varepsilon' &= \varepsilon_\infty + \frac{\varepsilon_s - \varepsilon_\infty}{1 + (\omega\tau)^2}, \\ \varepsilon'' &= \frac{\omega\tau\left(\varepsilon_s - \varepsilon_\infty\right)}{1 + (\omega\tau)^2}. \end{aligned} \tag{2.3}$$

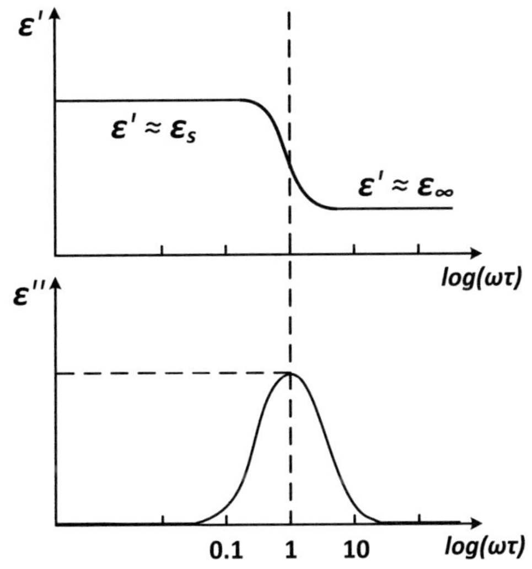

Figure 2.3. Amplitudes of ε' and ε'' at different frequencies [Met96].

According to above equations, the varying curves of ε' and ε'' are shown in figure 2.3.

Intuitively, the real amplitude ε' can be considered as the in-phase response with the external electric field $\overrightarrow{\mathbf{E}}$, which defines the amount of energy that can be stored within the dielectric and does not cause energy loss. The imaginary amplitude ε'' can be considered as the response that has $90\,^{\circ}$ phase shift with $\overrightarrow{\mathbf{E}}$, which determines the amount of energy dissipation from the external electric field and the heat generation within the dielectric. Correspondingly, the density of power dissipated into the dielectric due to the dipole relaxation is [Mer98]

$$p_{\mathrm{d}} = \frac{1}{2}\omega\varepsilon_0\varepsilon''(\omega)\left\|\overrightarrow{\mathbf{E}}\right\|^2, \tag{2.4}$$

where $\varepsilon_0 = 8.85\times10^{-12}\mathrm{F/m}$ is the permittivity of vacuum. It is evident from above equations as well as figure 2.3 that $f_0 = 1/\tau$ is the critical point where the dipole polarization fails to follow the direction change of $\overrightarrow{\mathbf{E}}$. Therefore it is more efficient to use microwaves with frequencies close to f_0 for heating. For different dielectric materials, the frequency f_0 is different. Many dielectric materials have f_0 locating between 0.8 GHz $\sim$ 3 GHz, and that is why the frequencies for industrial heating are 915 MHz and 2.45 GHz.

Besides dipole relaxation, there is another heat generation mechanism of microwave heating: ionic conduction. Ionic conduction is the movement of free ions or ionic species under the influence of external electric field, within a solid (such as structurally disordered crystalline solids) or aqueous solutions (such as solution of salts) [Tul07]. It is essentially similar with the electronic conduction of conductors, except the moving particles are ions instead of electrons. The energy is transformed from the electric field to the material and an amount of heat is generated. The ionic conduction loss is characterized by an equivalent dielectric conductivity σ_{d}, and the corresponding heating power density can be represented as [Met96]

$$p_{\mathrm{c}} = \frac{1}{2}\sigma_{\mathrm{d}} \left\| \overrightarrow{\mathbf{E}} \right\|^2 \tag{2.5}$$

Combining above two loss mechanisms, the overall microwave heating power density is expressed as [Met96]

$$\begin{aligned} p_{\mathrm{mw}} &= p_{\mathrm{d}} + p_{\mathrm{c}} = \frac{1}{2}\sigma_{\mathrm{e}}(\omega, T) \left\| \overrightarrow{\mathbf{E}} \right\|^2, \\ \sigma_{\mathrm{e}}(\omega, T) &:= \sigma_{\mathrm{d}} + \omega\varepsilon_0\varepsilon''(\omega, T), \end{aligned} \tag{2.6}$$

where $\sigma_{\mathrm{e}}(\omega, T)$ is the effective electrical conductivity. The total effective loss factor is defined as [Met96]

$$\varepsilon_{\mathrm{e}}''(\omega, T) := \frac{\sigma_{\mathrm{e}}(\omega, T)}{\omega\varepsilon_0} = \frac{\sigma_{\mathrm{d}}}{\omega\varepsilon_0} + \varepsilon''(\omega, T), \tag{2.7}$$

which represents the microwave energy absorbing ability of a dielectric material. The effective loss factor for fixed temperature is shown in figure 2.4, which gives a clear impression about the influences of different loss mechanism at different frequencies. In this dissertation, the effective electrical conductivity $\sigma_{\mathrm{e}}(\omega, T)$ is considered as a parameter $\sigma_{\mathrm{e}}(T)$ that relies on only the temperature T. Because it is assumed in the dissertation that the microwave frequency is fixed at 2.45 GHz.

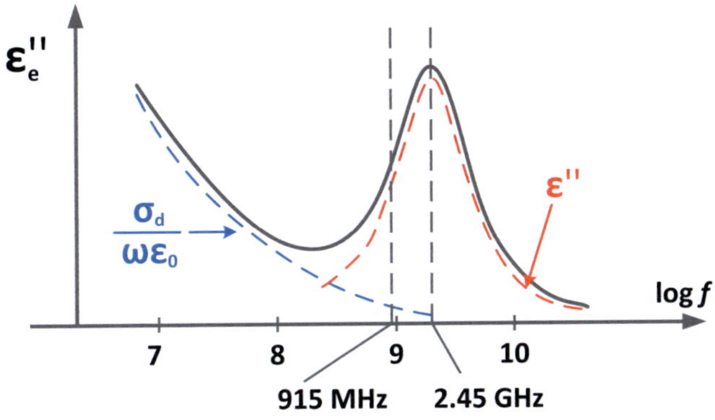

Figure 2.4. Effective loss factor due to the dipole relaxation and the ionic conduction losses [Met96].

2.2. Structure of HEPHAISTOS

HEPHAISTOS is a giant microwave heating system equipped with spatially distributed feeding sources (see figure 2.5). The size of a standard HEPHAISTOS module is about 180 cm × 155 cm × 110 cm, which is comparable to large industrial autoclaves (such as in figure 2.5b) and much larger than other microwave heating systems used anywhere else [lar]. More important, multiple HEPHAISTOS modules can be integrated into one large heating system, which is flexible to deal with large-scale workpieces.

Compared with other microwave heating systems, HEPHAISTOS is much more complex. It not only has a complicated hardware structure, but also consists of a sophisticated process control system and combined temperature measurement techniques. In general, the whole HEPHAISTOS system can be divided into three parts, indicated by three different colors in figure 2.6. These three parts are introduced in the following of this chapter.

(a) Photo of HEPHAISTOS.

(b) Autoclave in CFRP curing [aut].

Figure 2.5. Comparison between HEPHAISTOS (a) and an autoclave (b).

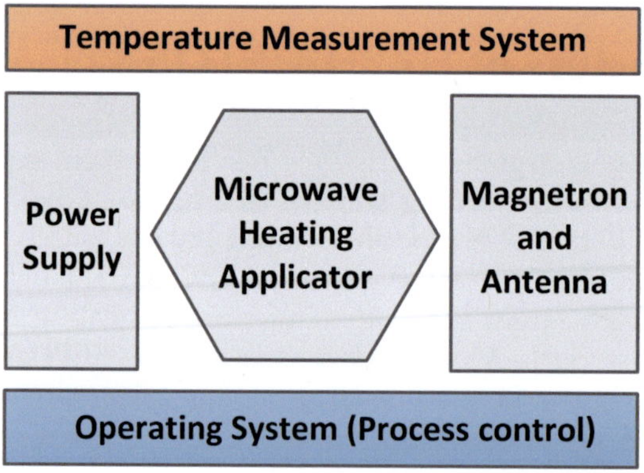

Figure 2.6. Schematic setup of HEPHAISTOS [Feh09].

2.2.1. Microwave Heating System

The structure of a typical microwave heating system is shown in figure 2.7. It consists of three parts, including a microwave generator, a waveguide and a resonant cavity. Since HEPHAISTOS is a distributed microwave heating system, it has multiple microwave generators and waveguides. Microwaves generated by each microwave generator will be transmitted through the corresponding waveguide, fed into the same resonant cavity, and finally absorbed by the heat load. The functions of individual parts are explained in the following.

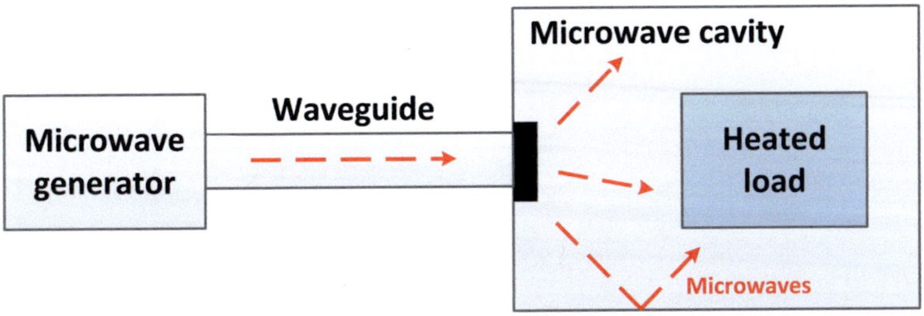

Figure 2.7. Structures of a normal microwave heating system.

Microwave Generator

A microwave generator is comprised of a power supply system and a microwave source. The power supply system provides the electrical energy to the microwave source, and the microwave source uses the electrical energy to generate microwaves and propagate them further to the succeeded equipments.

There are mainly two types of microwave sources, which are microwave tubes (vacuum devices) and solid-state devices. Microwave tubes are the most basic and fundamental microwave sources. There are many different types of microwave tubes, such as klystrons, magnetrons, gyrotrons and traveling-wave tubes [Jon98]. Microwave tubes have been widely used in various industrial applications, especially for those requiring high power or high frequencies [Poz09]. Compared with microwave tubes, solid-state devices normally have smaller size, lighter weight and smaller frequency shifts. The main limits of solid-state devices are that they could not generate microwaves of high power or high frequencies. Therefore solid-state devices are mainly applied for low to moderate powers at low to moderate frequencies. The average output power versus frequency performance of the microwave tubes and the solid-state devices is shown in figure 2.8.

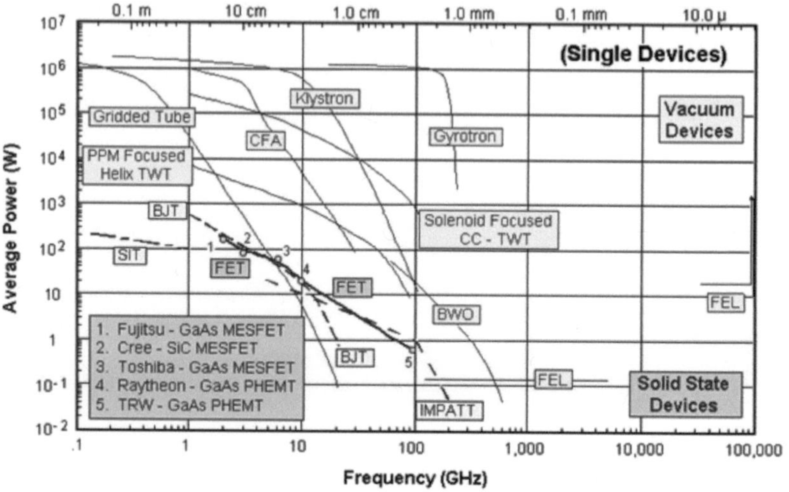

Figure 2.8. Power versus frequency performance of different solid-state devices and microwave tubes [Tre05].

In HEPHAISTOS, magnetron [Poz09] is selected as the microwave source, such as shown in figure 2.9. Magnetron is one of the most popular microwave tubes used in industrial applications. Its working principles are well explained in [Jon98] and [Gil11]. The reason why magnetron is used in HEPHAISTOS is due to considerations of size, cost, efficiency and most important, the ability to produce high power microwave (700 W to 2 kW).

These magnetrons are directly connected with a number of power supplies. In the old HEPHAISTOS cavity 3, power supplies with only pulse DC output were applied. All magnetrons had to work in the pulse mode (only ON and OFF), and different microwave power levels were only realized by adjusting the ON/OFF time ratio via the pulse width modulation (PWM) method [Bar01]. After the upgrade of the whole system, the new HEPHAISTOS cavity 3 are equipped with two different types of power supplies, such as shown in figure 2.10. Both of them are able to operate the magnetron in the continuous-wave (CW) mode, which means the microwave power can be adjusted continuously without PWM.

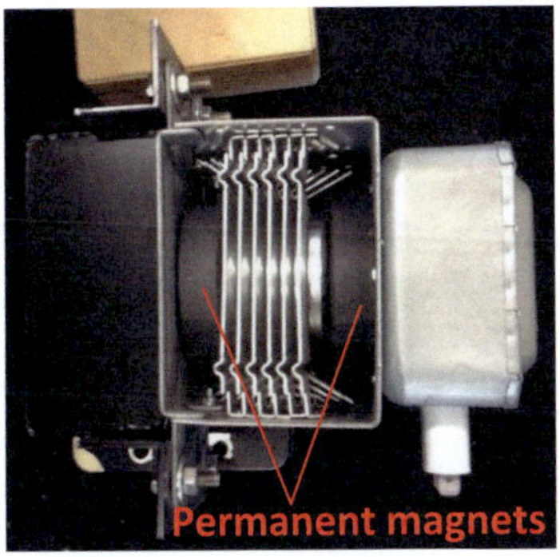

(a) Photo of a magnetron

(b) Cross-section of a magnetron

Figure 2.9. Photos of Magnetron used in practice (a) and its cross-section (b).

The power supplies in the first two modules of HEPHAISTOS cavity 3 are standard-ripple power supplies [rip], which are sufficient to fulfill all requirements of normal microwave heating. Another type of low-ripple power supplies is used in the third module, which can be used in applications requiring faster and more accurate output power.

(a) Standard-ripple power supply.

(b) Low-ripple power supply.

Figure 2.10. Power supplies used in the updated HEPHAISTOS cavity 3.

Waveguide

The role of waveguides is to transmit the electromagnetic wave from one point to another. In the microwave heating system, it is used to propagate microwaves from the generator to the microwave cavity. The waveguide is important for the whole microwave heating system, because it directly determines how much power can be fed to the microwave cavity. There are a variety of different types of waveguides, such as dielectric waveguides, coaxial cables, microstrips or hollow metal pipes [HB01]. Hollow metal pipes are the most common one used at microwave frequencies, such as in figure 2.11.

Figure 2.11. Hollow metal waveguides [web].

In HEPHAISTOS, there are two types of waveguides used. One is the WR-340 waveguide, which is a rectangular hollow metal waveguide used to transmit microwaves ranging from 2.20 GHz to 3.30 GHz [BBG+03] (as in figure 2.12a). The other one is the slotted waveguide, which acts as an antenna and omits the microwave to the external space [ST12] (such as shown in figure 2.12b). Each magnetron is directly mounted with a WR-340 waveguide, and every WR-340 waveguide is further connected to a slotted waveguide. The microwaves are transmitted from the microwave generator to the slotted waveguide via the WR-340 waveguide, and then directly radiated into the microwave cavity through the slots of the slotted waveguide. Details about the waveguides design refer to [Mer98] [HB01] and [Str07].

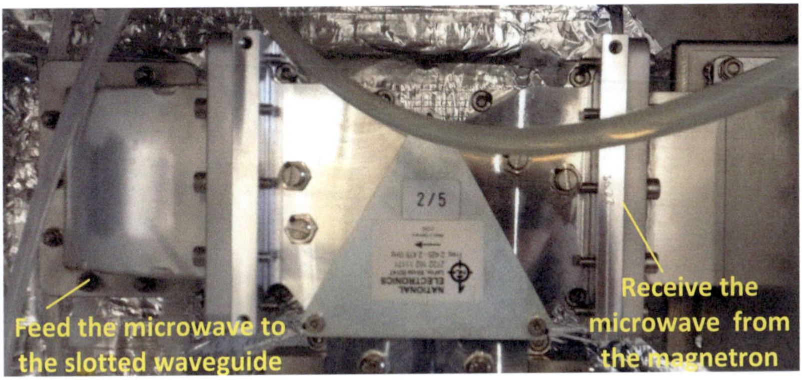

(a)

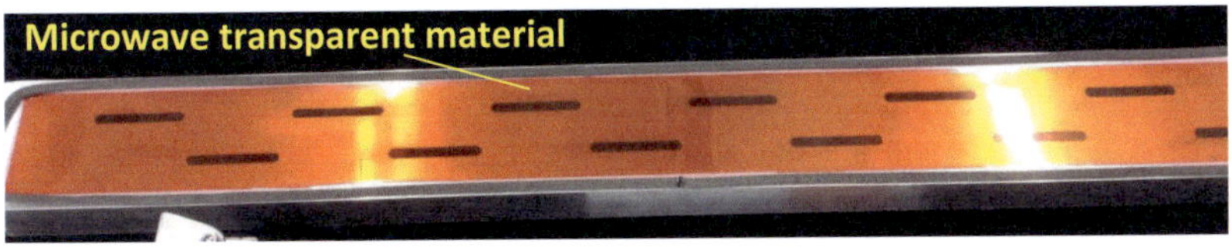

(b)

Figure 2.12. The WR-340 waveguide (a) and slotted waveguide (b) used in HEPHAISTOS

Microwave cavity

Microwave cavity is a type of resonator with a closed metal structure, which is the main part of a microwave heating system [Mer98]. Microwaves transmitted from waveguides will propagate through the cavity space, and they will also get reflected by the walls of the cavity, or get absorbed by the load that is put within the cavity. The microwave energy is gradually dissipated, and the load eventually gets heated. A microwave cavity could be constructed into different shapes, including rectangular, circular, hexagon (HEPHAISTOS), octagon [LLHG14] and others. The shape of a microwave cavity significantly influences the electromagnetic field modes that can exist within the cavity. According to the number of excited modes, the microwave cavity can be classified into the following two types.

- Single mode cavity

  The single mode cavity is designed to operate at a single resonant mode, which means the amplitude of this resonant frequency is significantly larger than that of the others. For a single mode cavity with the resonant frequency at f_0, its size is normally limited by the half wavelength $\lambda_0/2$ (λ_0 is the wavelength), e.g. 6.1 cm at 2.45 GHz. Because of the small size, single mode cavities are inappropriate to heat large-scale products, and not widely used in industrial microwave heating. But on the other hand, for the same power applied, the electric field strength and the corresponding microwave heating power density in a single mode cavity is much higher than a multimode cavity [MM83], therefore it is suitable to heat low loss dielectric materials. In addition, due to the narrow frequency band and simple geometry, the EM field distribution within a single mode cavity can be determined analytically, which gives direct guides of where and how to put the heated load to get the desired heating characteristics [FBLI+95].

- Multimode cavity

  Compared with the single mode cavity, the multimode cavity is much more complex. In general, the size of a multimode cavity is much larger than a single mode cavity, which gives the multimode cavity the ability to handle a wide range of workpieces with different sizes. More important, a large number of different modes can be excited within a multimode cavity, which leads to a more uniform EM field distribution and heating pattern than in a single mode cavity [HB01]. Due to these reasons, multimode cavities have been dominantly used in industrial microwave heating applications [Mer98]. Although multimode cavities are convenient to use, both design and analysis of multimode cavities are difficult.

A very important parameter for both single mode and multimode cavities is the quality factor Q, which indicates how much energy is dissipated compared to the total stored energy referring to the resonant frequency f_0, or equivalently, the 3 dB bandwidth [HB01] within a cavity regarding to this frequency. The quality factor Q of the mode n is defined by [Mer98]

$$\begin{aligned} Q_n &= 2\pi \frac{\text{total energy stored in mode } n}{\text{energy dissipated per cycle}} \\ &= 2\pi \frac{\text{total energy stored in mode } n \times \text{resonant frequency} f_n}{\text{power dissipated in mode } n}, \end{aligned} \tag{2.8}$$

or

$$Q_n \approx \frac{f_n}{\Delta f}, \tag{2.9}$$

where Δf is the 3 dB bandwidth defined as the frequency range with half power of the resonant frequency f_n.

Based on equation 2.8, the total quality factor Q of the whole cavity can be further divided as [Mer98],

$$\frac{1}{Q} = \frac{1}{Q_{\mathrm{L}}} + \frac{1}{Q_{\mathrm{W}}} + \frac{1}{Q_{\mathrm{D}}}. \tag{2.10}$$

where Q_{L} corresponds to the loss absorbed by the heated load, Q_{W} corresponds to the loss due to the dielectric properties of the wall, and Q_{D} consists of the diffraction loss leaking through unclosed parts of the cavity.

HEPHAISTOS is a giant multimode microwave cavity. From figure 2.5, it is clear that it has a very unique hexagonal shape. According to the simulation results in [Feh09], the electric energy density in the hexagonal shape applicator is much more homogeneous than densities of other shapes (including octagon, heptagon, cylindric and circular). In other words, the unique hexagonal design is able to reduce the EM field fluctuations and increase the EM field homogeneity over most of the cavity volume, which is also helpful to get a more uniform temperature distribution on the heated load [Feh09].

Unlike most of other microwave heating systems with only one source, in HEPHAISTOS there are more than one microwave heating sources per module. There are several advantages for using multiple feeding sources at different locations, e.g. the improvement of the EM field uniformity due to the cross coupling of multiple sources [Dat01]. But the most important one is that, as stated in [MM83], the implementation of multiple feeding sources creates the opportunity of adjusting the power so as to improve both the EM field and the temperature distributions, by the superposition of microwaves from different feeding sources.

For the old generation of HEPHAISTOS, each module has 12 microwave generators as feeding sources, and the total power of 12 microwave generators is 10 KW. The locations of each slotted waveguide are shown in figure 2.13 (from 1 to 12). In the upgraded HEPHAISTOS cavity 3, the number of microwave generators is reduced to 6 per module, and their locations are shown by the red labels in 2.13. With the new power supplies, each source is able to provide maximum 2 KW microwave power. Therefore the total power of one module is increased from 10 KW to 12 KW. This upgrade not only increases efficiency of each generator, but also reduces the area of unclosed parts of the cavity, which lowers the quality factor Q and further enhances the efficiency of the whole cavity.

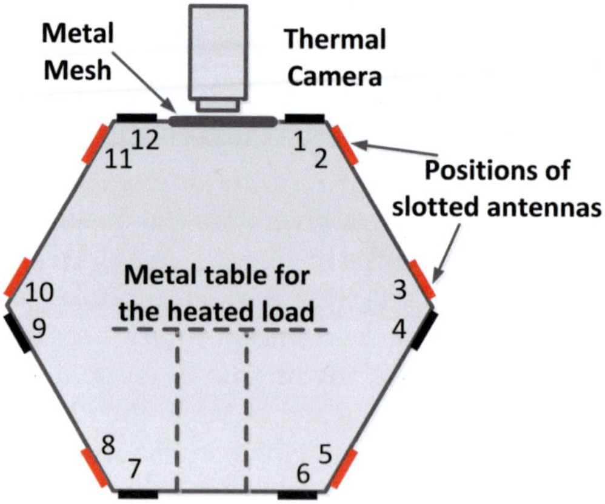

Figure 2.13. Cross section structure of HEPHAISTOS.

2.2.2. Process Control System

The process control system of HEPHAISTOS mainly contains two parts, including a LabVIEW [lab] [TK06] based control interface and another MATLAB [mat] [EEE97] based temperature control system. The whole diagram of the process control system is shown in figure 2.14.

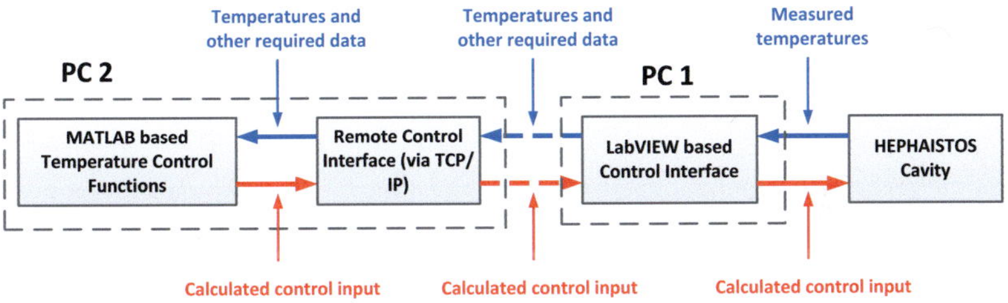

Figure 2.14. Diagram of the process control system.

The LabVIEW based control interface directly communicates with the HEPHAISTOS cavity using INTERBUS [FT88] (used in the current HEPHAISTOS cavity) or PROFINET [JF05] (used in the Hybrid HEPHAISTOS cavity [Gmb14]) serial protocol. It receives the measured temperature values from the cavity (or from the infrared camera in the other case), and sends different parameter settings and commands to the cavity, like the control input, heating time, and target temperature value. Simple temperature control schemes, such as manual control (manually set power of each source) and PID control, can be directly realized in the LabVIEW using its VIs [lab] and toolboxes. The front panel of the interface is demonstrated in figure 2.15.

For more complicated temperature control methods, such as all the adaptive and the intelligent control methods aforementioned, they have to be called from a remote MATLAB based control system. That is because LabVIEW as a graphical programming language is not efficient to deal with high dimensional matrix computation tasks, which are easily to be done in MATLAB. These two parts are running in different PCs in order to guarantee the sufficient computation resources for both of them. During the heating process, they communicate through a remote control interface, which is shown in figure 2.16.

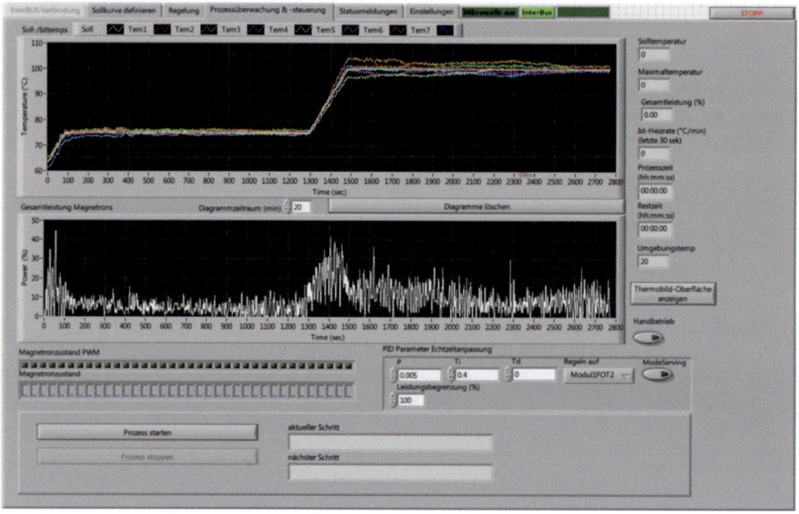

Figure 2.15. Labview based control interface.

Figure 2.16. Remote control interface of HEPHAISTOS.

The remote control interface functions like a transfer station. It receives different types of data, including the temperature values, prediction length (used in MPC) and other required information, from the LabVIEW control interface, and send them to the MATLAB control system. After the new control input is calculated by the MATLAB control system, the remote interface picks it up and further sends it to the LabVIEW control interface. The communication between the remote interface and the LabVIEW control interface is done through the TCP/IP protocol.

2.2.3. Temperature Measurement Approaches

In HEPHAISTOS, there are three different temperature measurement approaches, including the thermocouple, fiber optic sensor and infrared camera. The infrared camera is put right above the metallic table, which is used to get the whole temperature profile of the heated workpiece during the heating process. Both thermocouples and fiber optic sensors can be fixed at any locations using microwave transparent tapes, to measure the surface temperature of the workpiece.

Different temperature measurement approaches have different performance in practical experiments. For thermocouples, their presence in the microwave cavity significantly influence the EM field, leading to thermal instability and microwave breakdown [PCB+01]. Even shielded with aluminium or copper backed tape, the shielding itself will be heated which also affects the measurement accuracy of thermocouples. Compared with thermocouples, the fiber optic sensor is much more convenient that can be directly used without any shielding or other processing. But the main problem is its large time delay during the temperature measurement.

The figure 2.17 demonstrates the time delay occurred in temperature measurement using fiber optic sensors. After the microwave sources are switched on, the temperatures measured by fiber optic sensors stayed constant for quite a period. The time between switching on heating power and temperature changes is around 8 s to 10 s. What is even worse is that this time delay value is not fixed. It differs from time to time and from sensor to sensor. This time delay brings extra difficulties to both the control of the system and the temperature measurement. There is no way to eliminate or reduce its influences from neither hardware nor modeling aspects.

The above mentioned problems of thermocouples and fiber optic sensors can be easily overcome by the infrared camera. The temperature measurement using the infrared camera is contactless, and it can be directly implemented without any additional processing. As also shown in figure 2.17, the time delay of the infrared camera is so small that can be neglected in practice. The only difficulty regarding the infrared camera is the temperature mismatch caused by the metallic mesh that is put in between the camera and the cavity.

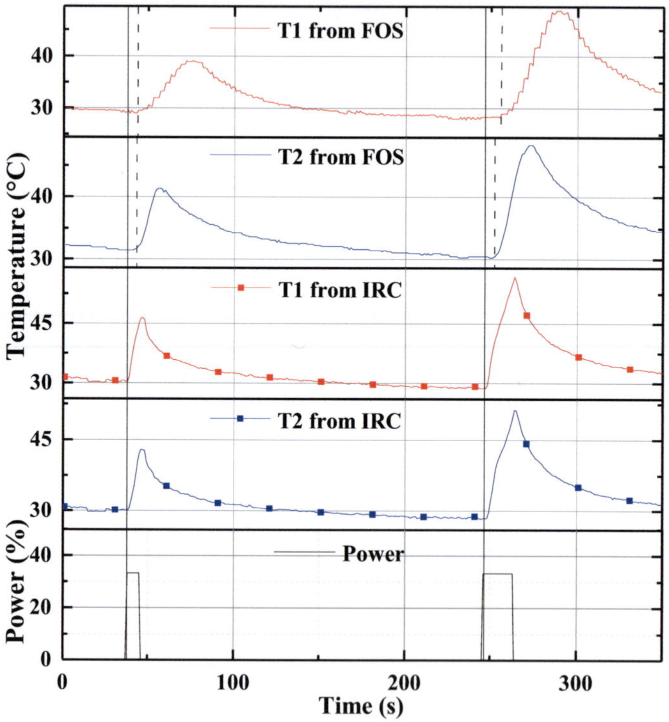

Figure 2.17. Comparison of measurement delays using the fiber optic sensor (FOS) and the infrared camera (IRC).

In principle, the metallic mesh is designed to block the microwave and keep the safety of the infrared camera. But during the heating process, this mesh absorbs parts of the infrared signals from the workpiece, and meanwhile it also emits a certain amount of infrared signals. The infrared signals received by the infrared camera are partly from the workpiece and partly from the mesh itself. As a result, the temperature directly measured by the infrared camera is different from the real temperature of the workpiece. This difference is proportional to the temperature difference between the workpiece and the mesh. Therefore when the workpiece is heated to a high temperature and the mesh stays cool, the temperature values measured by the infrared camera will be significantly differ from the real temperatures. In order to get the accurate temperature measurements, this influence has to be compensated.

According to the above principles, the mesh can be considered as an infrared window [Gru03]. Once the emissivity of the workpiece, the transmittance of the mesh and the temperature of the mesh are known, the measured temperature can be compensated using the method as described in [Mad04]. In HEPHAISTOS, the compensation functions are included and applied using the FLIR ThermoVision LabVIEW Tool-kit [fli].

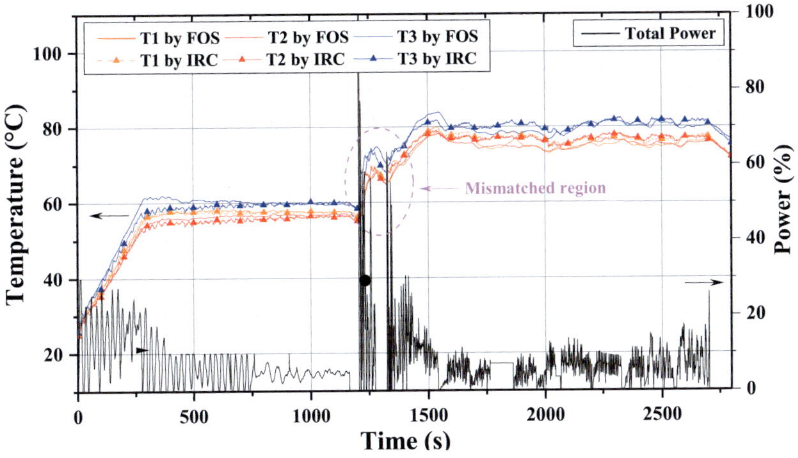

(a)

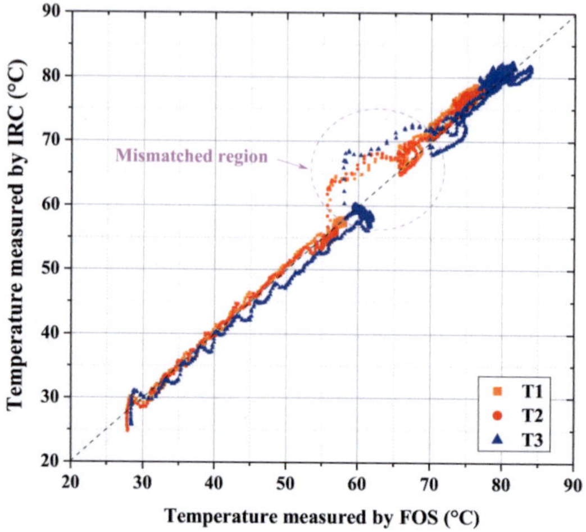

(b)

Figure 2.18. Temperature comparison between FOS and IRC (after compensation).

After the compensation, the temperature measured by the infrared camera is further verified. Temperatures of three different locations on the surface of a workpiece were measured by both fiber optice sensors and the infrared camera. The corresponding temperature curves and fitting results are shown in figure 2.18.

From the result in figure 2.18, it is clear that after the compensation the temperatures measured by the infrared camera are close to the temperatures measured by the fiber optic sensors. The differences between them are limited within $\pm 2°C$ except the mismatched region, which is labeled in both figures in 2.18. This region is caused by the fact that when the temperatures have large changes (see figure 2.18a), the response of the infrared camera is much faster than the response of fiber optic sensors. In other words, this mismatched region is generated because of the time delay of fiber optic sensors. Due to the fast and accurate temperature measurement results, the infrared camera is mainly used in the following experiments.

3. Modeling Microwave Heating

Based on the introduction to microwave heating and HEPHAISTOS in the previous chapter, the mathematical models of the microwave heating process in HEPHAISTOS are built in this chapter. A brief introduction of mathematical modeling is given first, followed by two different modeling approaches, the grey-box modeling and the black-box modeling. For each modeling approach, the corresponding system identification (SID) algorithms, which are used to estimate the unknown parameters in the model, are also explained. Both approaches have been implemented, to identify the most suitable modeling scheme for the temperature control system of HEPHAISTOS. The performance comparison can be found in chapter 5.

In this dissertation, we are focusing on controlling and improving the surface temperature distribution of a rectangular dielectric foil setup, such as shown in figure 3.1.

Two assumptions have been made:

1. The load receives most of electromagnetic energy (heating energy) from the top surface.
2. The thickness of the load is much smaller than the wavelength of the microwave ($\ll$ 12.5 cm).

These two assumptions guarantee that the surface temperature distribution of the load is equivalent to the temperature distribution within the load (similar assumptions are used in [CRB08] and [BRS01]). In other words, the temperature difference in the vertical direction (z-direction as in figure 3.1) is sufficiently small. Thus the heat conduction on the vertical direction can be neglected, which simplifies the entire model and following derivations. If the surface temperature distribution is controlled and improved, the inner temperature distribution can be controlled and improved consequently.

Besides above two assumptions, it also should be noted that the heat conduction effect caused by the tool (in figure 3.1) is not included in the modeling process. In many cases the heat conduction effect caused by the tool has a significant influence to the final temperature distribution of the load. It directly determines the accuracy of the control model (constructed in this chapter), even the controllability [Dor95] of the entire microwave heating system. However, many practical factors have to be taken into account to fully analyze this influence and there is no appropriate way to include it in the control model. Influences of tools made of different materials are compared and explained in chapter 5.

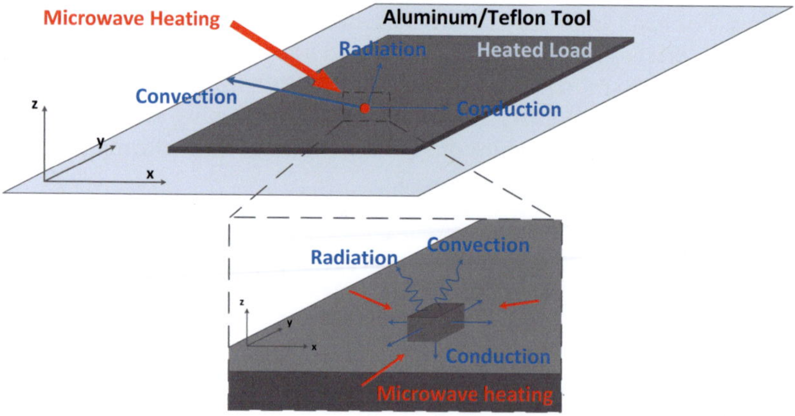

Figure 3.1. Sketch of microwave heating setup.

3.1. Mathematical Modeling

In general, the objective of mathematical modeling is to identify accurate and comprehensive mathematical concepts or equations to describe the dynamics of a system. Based on the mathematical model, the future behaviors of a system can be predicted, and different properties of the system can be analyzed. As mentioned in chapter 1, a good mathematical model should be able to both describe the main characteristics of a system, but also keep the complexity at a low level. Depending on the nature of different systems, the corresponding mathematical models are distinguished into static, dynamic, explicit, implicit, deterministic, stochastic or other models. Essentially there are three main categories of mathematical modeling approaches.

Black-box modeling

In black-box modeling [Lju97], as the name suggests, the investigated system is treated as a completely unknown object. No physical or any other kind of prior knowledge is available, and the whole system only consists of three parts: the input, the output and the transfer relationship (function) such as in figure 3.2. In a black-box, the choice of input and output components is based on observations and requirements accordingly. The critical task is to select an appropriate model and function structure [JHB+95]. For example, for a system with multiple input and output variables, the state-space model [Lju97] is preferable to the transfer function formed model [Lju97]. The principle of selecting the right model structure is to guarantee that the main features of the system can be reflected by the model. Besides, the number of assumptions should be as few as possible. Normally the model structure is chosen on the basis of experimental data or empirical knowledge [SZL+95].

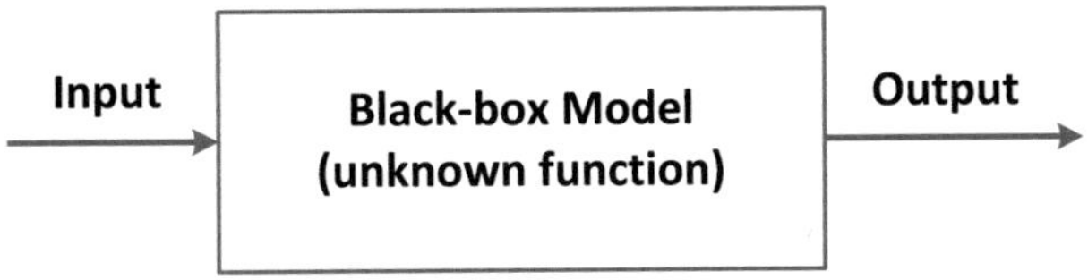

Figure 3.2. Diagram of black-box modeling.

White-box modeling

The white-box modeling is the opposite of black-box modeling. In a white-box system, all properties and details of the system are well defined by first principles, and system parameters are also perfectly known [JHB+95]. Although white-box modeling is theoretically clear, it is rarely used to solve real problems. In many cases the white-box model is too complicated or even impossible to implement and solve within a given time.

Grey-box modeling

Grey-box modeling is a combination of the black-box and the white-box modeling, where not all but a certain part of the physical insight is available [JHB+95]. The model structure can be directly determined by prior knowledge and physical insight, but many system parameters still remain unknown. These unknown parameters have to be estimated using experimental data via either online or offline system identification (SID) methods. In many cases, the grey-box model is sufficiently accurate to interpret the relationship between inputs and outputs of the system, without going into unneeded details.

A pure white-box system seldom exists in practice, therefore for real problems, the black-box modeling and the grey-box modeling are mainly used. Both of them have their advantages and disadvantages. Black-box modeling is essentially more flexible and it can be applied to all kinds of systems. In theory, black-box modeling can reach arbitrarily high accuracy given an adequate amount of experimental data [JHB+95]. No other knowledge or additional assumptions are needed. But on the other side, the performance of a black-box model depends on many factors, which means the model itself has to be tuned and optimized to get an acceptable performance.

For instance, as one of the most popular black-box modeling methods, the neural network (NN) method has been implemented in many areas. The performance of a neural network is determined by several components, such as the number of hidden layers or the number of nodes in each hidden layers. There is no sophisticated guide how to choose these elements. All elements have to be compared and adjusted according to the specific situation, which could take a long time and require extra efforts. Although black-box modeling has great potential to achieve a good performance, it is normally considered as the last resort and only utilized when no alternative is available [Lju97].

In comparison with black-box modeling, the number of unknown parameters and the amount of experimental data required in grey-box modeling are generally less. Moreover, the first principles used in grey-box modeling also give certain physical interpretations to the parameters being estimated, which double-check the estimation accuracy and help to understand the internal physical properties of the system. However, in many cases the first principles are still too complicated to be directly applied or not complete to cover all dynamics of the original system.

In order to identify the most suitable model for the microwave heating process and make HEPHAISTOS adaptive to different heating scenarios, both grey-box and black-box modeling approaches are applied in this dissertation.

3.2. Grey-box Modeling

There are mainly three procedures to apply the grey-box modeling approach. The first step is to identify the physical principles that govern the entire energy exchange process. If necessary, certain simplification and approximation have to be implemented to transfer the original principles into controllable forms. The last step is to organize and discretize the resulted models, in order to make them suitable for the following controller design.

3.2.1. First Principles

For any unit cell on the surface of the heated load such as in figure 3.1 with size of $l \times l \times l$ (e.g. l =1 cm), the complete energy (temperature) changing equation can be derived according to the law of conservation of energy [LNS04], such as indicated by the equation

$$\begin{aligned} \text{Net amount of absorbed power } P_{\mathrm{ab}} &= \text{Dissipated power due to heat convection } P_{\mathrm{cv}} \\ &+ \text{Dissipated power due to heat radiation } P_{\mathrm{rd}} \\ &+ \text{Dissipated/heating power due to heat conduction } P_{\mathrm{cd}} \\ &+ \text{Microwave heating power } P_{\mathrm{mw}}. \end{aligned} \tag{3.1}$$

Each element in the above equation can be calculated as following.

- Net amount of absorbed power

  The net amount of absorbed power P_{ab} within this unit cell is represented by [LNS04]

$$P_{\mathrm{ab}} = \rho c_{\mathrm{p}} \frac{\mathrm{d}T}{\mathrm{d}t} \cdot l^3, \tag{3.2}$$

  with

  ρ : mass density, $\mathrm{kg/m^3}$
  c_{p} : specific heat per unit mass at constant pressure, $\mathrm{J/(K \cdot kg)}$

- Dissipated power due to heat convection

  The power flux density of convective heat transfer p_{cv} ($\mathrm{W/m^2}$) is given by Newton's Law of cooling [Win99], such as

$$p_{\mathrm{cv}} = -h(T - T_{\mathrm{a}}), \tag{3.3}$$

  where $T - T_{\mathrm{a}}$ is the temperature difference between the load and the surrounding air flow (T_{a}). The parameter h is the convection heat transfer coefficient ($\mathrm{W/(m^2 \cdot K)}$), which is determined by the geometry of the load, the velocity of the air flow, the angel between the surface and air flow and other elements. The negative sign indicates the heat is flowing from the load to the surrounding air. According to equation 3.3, the dissipated power due to the heat convection P_{cv} can be calculated as

$$P_{\mathrm{cv}} = p_{\mathrm{cv}} \cdot l^2 = -h(T - T_{\mathrm{a}}) \cdot l^2, \tag{3.4}$$

where l^2 denotes the area of the contacting surface between the cell and the surrounding air.

- Dissipated power due to heat radiation

  The maximum radiation flux density p_{rd} from a black surface is defined by the Stefan-Boltzmann Law [LNS04], which is

  $$p_{\mathrm{rd}} = -\sigma' T^4, \tag{3.5}$$

  where $\sigma' = 5.669 \times 10^{-8} \mathrm{W}/\left(\mathrm{m}^2 \cdot \mathrm{K}^4\right)$ is the Stefan-Boltzmann constant. But in practice for a setup as in figure 3.1 with a non-black surface, where the load is completely surrounded by a much larger volume of air, the flux density p_{rd} can be also calculated by the equation [LNS04]

  $$p_{\mathrm{rd}} = -\varrho\sigma'(T^4 - T_{\mathrm{a}}^4), \tag{3.6}$$

  where ϱ is the emissivity of the surface ($0 \leq \varrho \leq 1$, 1 for black surface), indicating the radiation ability of the material [Sie01]. The dissipated power due to the heat radiation P_{rd} can be similarly represented as

  $$P_{\mathrm{rd}} = p_{\mathrm{rd}} \cdot l^2 = -\varrho\sigma'(T^4 - T_{\mathrm{a}}^4) \cdot l^2. \tag{3.7}$$

- Dissipated/heating power due to heat conduction

  According to aforementioned assumptions 1 and 2, the provided or dissipated power by the heat conduction can be decomposed into the components in the x- and y-directions [LNS04], such as

  $$P_{\mathrm{cd}} = \frac{\partial}{\partial x}\left(\kappa_x l^2 \frac{\partial T}{\partial x}\right) \cdot l + \frac{\partial}{\partial y}\left(\kappa_y l^2 \frac{\partial T}{\partial y}\right) \cdot l, \tag{3.8}$$

  where κ_x and κ_y indicate thermal conductivities in the x- or y-direction, respectively. If the heated load has the same thermal conductivity in different directions like $\kappa_x = \kappa_y = \kappa$, then the above term can be further written as

$$\begin{aligned} P_{\mathrm{cd}} &= \frac{\partial}{\partial x}\left(\kappa_x l^2 \frac{\partial T}{\partial x}\right) \cdot l + \frac{\partial}{\partial y}\left(\kappa_y l^2 \frac{\partial T}{\partial y}\right) \cdot l \\ &= \frac{\partial}{\partial x}\left(\kappa \frac{\partial T}{\partial x}\right) \cdot l^3 + \frac{\partial}{\partial y}\left(\kappa \frac{\partial T}{\partial y}\right) \cdot l^3 \\ &= \nabla \cdot (\kappa \nabla T) \cdot l^3, \end{aligned} \tag{3.9}$$

where ∇ is the two-dimension differential operator. If the sign of $\nabla \cdot (\kappa \nabla T)$ is positive, the energy goes from other parts of the load to this cell, and vice versa.

- Microwave heating power

According to equation 2.6, the microwave heating power density is written as

$$p_{\mathrm{mw}} = \frac{1}{2}\sigma_{\mathrm{e}}(T) \left\| \vec{\mathbf{E}} \right\|^2, \tag{3.10}$$

where $\sigma_{\mathrm{e}}(T)$ is the electrical conductivity of the heated load at temperature T (equation 2.6). The electric field strength as well as the entire EM field distribution is determined by the well-known Maxwell's equations [Mer98] [Y$^+$66]. Unlike numerical simulation approaches, the exact solutions of the Maxwell's equations are not required for the temperature control system. Instead the microwave heating power densities are approximated and estimated using different system identification methods that will be introduced later this chapter. For a small cell as defined in figure 3.1, it is assumed that the electric field is constant within the cell, and correspondingly the microwave heating power is given by

$$P_{\mathrm{mw}} = p_{\mathrm{mw}} \cdot l^3. \tag{3.11}$$

Replacing the corresponding terms in equation 3.1 by above expressions, there is the complete heat transfer equation for any location (cubic cell) on the surface of the heated load

$$\overbrace{\rho c_{\mathrm{p}} \frac{\mathrm{d}T}{\mathrm{d}t} \cdot l^3}^{P_{\mathrm{ab}}} = \overbrace{\nabla \cdot (\kappa \nabla T) \cdot l^3}^{P_{\mathrm{cd}}} \overbrace{- h(T - T_{\mathrm{a}}) \cdot l^2}^{P_{\mathrm{cv}}} \overbrace{- \varrho \sigma' (T^4 - T_{\mathrm{a}}^4) \cdot l^2}^{P_{\mathrm{rd}}} \\ \underbrace{+ \frac{1}{2} \sigma_{\mathrm{e}}(T) \left\| \vec{\mathbf{E}} \right\|^2 \cdot l^3}_{P_{\mathrm{mw}}} . \tag{3.12}$$

It should be noted that in the above equation the power absorption or dissipation is already reflected by the sign of individual terms.

3.2.2. Approximation and Simplification

The above equation 3.12 is a complicated parabolic partial differential equation (PDE). This equation can be only solved in special cases like the one-dimensional scenario. For a microwave heating system like HEPHAISTOS, where the electromagnetic field distribution is complex and varying all the time during the heating, a general solution is impossible to be obtained. In order to make equation 3.12 suitable for a control system, approximation and simplification have to be done here, regarding different terms on the right-hand side.

Approximations of the thermodynamic terms

From equation 3.12, the overall dissipated power can be deduced, such as

$$\begin{aligned} P_{\mathrm{diss}} &= P_{\mathrm{cd}} + P_{\mathrm{cv}} + P_{\mathrm{rd}} \\ &= \nabla \cdot (\kappa \nabla T) \cdot l^3 - h(T - T_{\mathrm{a}}) \cdot l^2 - \varrho \sigma' (T^4 - T_{\mathrm{a}}^4) \cdot l^2, \end{aligned} \tag{3.13}$$

where the three terms on the right-hand side correspond to the power of conduction, convection and radiation, respectively. A very effective scheme to approximate this heat loss part is to ignore the rate of conduction and only keep the rate of convection and radiation, because the effect of conduction is quantitatively much smaller than the effect of the convection or the radiation.

For instance, a rectangular dielectric foil is heated to 350 K, and the temperature of surrounding air is 300 K. For the dielectric materials, it is assumed the corresponding heat conductivities are below 10 $\mathrm{W/(m \cdot K)}$. For example, the conductivity of silicon rubber (the material that is mostly used in our experiments) is between 0.2 $\mathrm{W/(m \cdot K)}$ to 1.3 $\mathrm{W/(m \cdot K)}$ [Sil12], and the conductivity of CFRP using epoxy matrices is 5 $\mathrm{W/(m \cdot K)}$ to 7 $\mathrm{W/(m \cdot K)}$ in plane [Tia11]. The second spatial derivative of the temperature is difficult to be defined. Based on the moderately homogeneous EM field distribution within HEPHAISTOS, it is feasible to use $\pm 0.1\mathrm{K/cm^2}$ as the second spatial derivatives. Other parameters are assumed as

$$\varrho = 0.9, \quad h = 20\,\mathrm{W/(m^2 \cdot K)},$$

Using above values, for a simple cell on the surface (1 cm $\times$ 1 cm $\times$ 1 mm) the rates of thermal conduction, convection and radiation can be calculated as

$$\begin{aligned} \left|P_{\mathrm{cd}}\right| &< 0.01\,\mathrm{W}, \\ P_{\mathrm{cv}} &= -0.1\,\mathrm{W}, \\ P_{\mathrm{rd}} &= -0.035\,\mathrm{W}. \end{aligned} \tag{3.14}$$

Above results show that the dissipated power of either thermal convection or thermal radiation is much larger than that of thermal conduction, therefore the thermal conduction term in equation 3.13 can be omitted. Furthermore, the dissipated power of both thermal convection and thermal radiation are proportional to the temperature difference $T - T_{\mathrm{a}}$ and the area of the unit cell l^2. In this case, the overall dissipated power is rewritten as

$$\begin{aligned} P_{\mathrm{diss}} &\approx P_{\mathrm{cv}} + P_{\mathrm{rd}} \\ &\approx -\tilde{h}(T)\,(T - T_{\mathrm{a}}) \cdot l^2, \\ \tilde{h}(T) &:= h + \varrho\sigma'(T^2 + T_{\mathrm{a}}^2) \cdot (T + T_{\mathrm{a}}), \end{aligned} \tag{3.15}$$

where $\tilde{h}(T)$ is defined as the overall power dissipation coefficient that contains effects of both thermal convection and thermal radiation. Substituting the above equation 3.15 into equation 3.12, the resulted system models has a similar ARMAX form like the models used in [RCVI99] and [SBA00].

Approximations of the microwave heating term

Compared with the heat loss part, the approximation of the microwave heating term is more difficult, because of the complex microwave heating power superpositions. In general, there are two types of microwave power superposition principles. The first power superposition principle corresponds to the situation when the EM fields from different sources are incoherent. In this case, the instantaneous microwave heating power of each source is varying, but the average heating power is constant. Therefore the superposed heating power is equivalent to the addition of heating power of individual sources.

For instance, assuming there are two microwave heating sources, the electric fields from two sources at the same location n are $\overrightarrow{\mathbf{E}}^{n,1}$ and $\overrightarrow{\mathbf{E}}^{n,2}$, respectively. The heating power from the two sources are represented by

$$P_{\text{mw}}^{n,1} \propto \left\|\overrightarrow{\mathbf{E}}^{n,1}\right\|^2, \qquad P_{\text{mw}}^{n,2} \propto \left\|\overrightarrow{\mathbf{E}}^{n,2}\right\|^2.$$

Then the superposed heating power [BPJD99] is

$$P_{\text{mw}}^{n} = P_{\text{mw}}^{n,1} + P_{\text{mw}}^{n,2} \propto \left\|\overrightarrow{\mathbf{E}}^{n,1}\right\|^2 + \left\|\overrightarrow{\mathbf{E}}^{n,2}\right\|^2. \tag{3.16}$$

For HEPHAISTOS, considering multiple heating sources and controllable feeding power, the total effective microwave heating power at the location n can be written as

$$P_{\text{mw}}^{n} = \sum_{m=1}^{M} u_m P_{\text{mw}}^{n,m}, \tag{3.17}$$

where u_m $(1 \leq m \leq M)$ stands for the portion of heating power from the m-th feeding source and $P_{\text{mw}}^{n,m}$ is the maximum heating power from the m-th source to the n-th location. The above equation 3.17 is rewritten in a matrix multiplication form as

$$P_{\text{mw}}^{n} = \mathbf{P}_{\text{mw}}^{n}\mathbf{U}, \tag{3.18}$$

with

$$\mathbf{U} := [u_1,\, u_2,\, \ldots,\, u_M]^{\mathrm{T}},$$
$$\mathbf{P}_{\text{mw}}^{n} := [\,P_{\text{mw}}^{n,1},\, P_{\text{mw}}^{n,2},\, \ldots,\, P_{\text{mw}}^{n,M}\,],$$

where **U** is the control input vector that is used to adjust the microwave heating power and corresponding heating rate.

The scalar superposition rule is simple to implement and control, which is the main approximation approach used in former researches about temperature controller design of microwave heating, such as [RMTV85] and [SBA00]. Since the above expression is a linear superposition, it is named as the linear microwave heating power approximation.

Besides the scalar addition rule, there is another principle of heating power superposition, which deals with the situation when all EM fields from different sources are coherent. When the EM fields have the same frequency and constant phase differences between each other, the microwave heating power of different sources superposes following the rule of vector addition. Using the same example as in equation 3.16, the superposed electric field at the location n can be represented by the equation

$$\overrightarrow{\mathbf{E}}^{n} = \overrightarrow{\mathbf{E}}^{n,1} + \overrightarrow{\mathbf{E}}^{n,2},$$

and the corresponding total heating power [BPJD99] is

$$P_{\mathrm{mw}}^{n} \propto \left\|\overrightarrow{\mathbf{E}}^{n}\right\|^{2} = \left\|\overrightarrow{\mathbf{E}}^{n,1}\right\|^{2} + \left\|\overrightarrow{\mathbf{E}}^{n,2}\right\|^{2} + 2\cdot\left\|\overrightarrow{\mathbf{E}}^{n,1}\right\|\cdot\left\|\overrightarrow{\mathbf{E}}^{n,2}\right\|\cos\Phi, \tag{3.19}$$

where Φ is the phase difference between $\overrightarrow{\mathbf{E}}^{n,1}$ and $\overrightarrow{\mathbf{E}}^{n,2}$. While $\cos\Phi \geq 0$, the superposition is constructive; otherwise while $\cos\Phi < 0$, the superposition is destructive.

Equation 3.19 can be extended to a more general case like HEPHAISTOS. Assuming there are M microwave heating sources, for each heating source m, its radiated electric field at the location n is denoted by $\overrightarrow{\mathbf{E}}^{n,m}$, which is defined as

$$\overrightarrow{\mathbf{E}}^{n,m} := \mathrm{E}_{x}^{n,m}\,\overrightarrow{\mathbf{e}_{x}} + \mathrm{E}_{y}^{n,m}\,\overrightarrow{\mathbf{e}_{y}} + \mathrm{E}_{z}^{n,m}\,\overrightarrow{\mathbf{e}_{z}}, \tag{3.20}$$

where $\mathrm{E}_{x}^{n,m}$ ($\mathrm{E}_{y}^{n,m}$, $\mathrm{E}_{z}^{n,m}$) and $\overrightarrow{\mathrm{e}_{x}}$ ($\overrightarrow{\mathrm{e}_{y}}$, $\overrightarrow{\mathrm{e}_{z}}$) are the amplitude of electric field component and unit vector in the x (y, z) -direction, respectively.

To make the electric field easier to use in the following derivations, it is rewritten in a vector form, such as

$$\overrightarrow{\mathbf{E}}^{n,m} = \mathbf{E}^{n,m} := \begin{bmatrix} \mathrm{E}_x^{n,m} \\ \mathrm{E}_y^{n,m} \\ \mathrm{E}_z^{n,m} \end{bmatrix}, \tag{3.21}$$

with

$$\overrightarrow{\mathbf{e}_x} = \mathbf{e}_x := \begin{bmatrix} 1 \\ 0 \\ 0 \end{bmatrix}, \quad \overrightarrow{\mathbf{e}_y} = \mathbf{e}_y := \begin{bmatrix} 0 \\ 1 \\ 0 \end{bmatrix}, \quad \overrightarrow{\mathbf{e}_z} = \mathbf{e}_z := \begin{bmatrix} 0 \\ 0 \\ 1 \end{bmatrix}.$$

Therefore the overall superposed electric field at the location $\overrightarrow{\mathbf{E}}^n$ can be replaced by the vector $\mathbf{E}^n$, such as

$$\mathbf{E}^n = \sum_{m=1}^{M} v_m \cdot \mathbf{E}^{n,m} = \begin{bmatrix} \mathrm{E}_x^{n} \\ \mathrm{E}_y^{n} \\ \mathrm{E}_z^{n} \end{bmatrix}, \tag{3.22}$$

where

$$\begin{aligned} \mathrm{E}_x^n &= v_1 \mathrm{E}_x^{n,1} + v_2 \mathrm{E}_x^{n,2} + \cdots + v_M \mathrm{E}_x^{n,M}, \\ \mathrm{E}_y^n &= v_1 \mathrm{E}_y^{n,1} + v_2 \mathrm{E}_y^{n,2} + \cdots + v_M \mathrm{E}_y^{n,M}, \\ \mathrm{E}_z^n &= v_1 \mathrm{E}_z^{n,1} + v_2 \mathrm{E}_z^{n,2} + \cdots + v_M \mathrm{E}_z^{n,M}, \end{aligned} \tag{3.23}$$

and v_m $(1 \leq m \leq M)$ is the portion of amplitude of the electric field from the m-th feeding source.

Defining the new vectors as

$$\begin{aligned} \mathbf{V} &= \left[v_1,\, v_2,\, \ldots,\, v_M \right]^{\mathrm{T}}, \\ \mathbf{E}_x^n &= \left[\mathrm{E}_x^{n,1},\, \mathrm{E}_x^{n,2},\, \ldots,\, \mathrm{E}_x^{n,M} \right], \\ \mathbf{E}_y^n &= \left[\mathrm{E}_y^{n,1},\, \mathrm{E}_y^{n,2},\, \ldots,\, \mathrm{E}_y^{n,M} \right], \\ \mathbf{E}_z^n &= \left[\mathrm{E}_z^{n,1},\, \mathrm{E}_z^{n,2},\, \ldots,\, \mathrm{E}_z^{n,M} \right], \end{aligned} \tag{3.24}$$

the components in the vector $\mathbf{E}^n$ can be expressed as

$$\mathrm{E}_x^n = \mathbf{E}_x^n \mathbf{V}, \quad \mathrm{E}_y^n = \mathbf{E}_y^n \mathbf{V}, \quad \mathrm{E}_z^n = \mathbf{E}_z^n \mathbf{V}. \tag{3.25}$$

Then there is

$$
\begin{aligned}
\left\|\vec{\mathbf{E}}^n\right\|^2 &= \mathbf{E}^{n\mathrm{T}}\mathbf{E}^n \\
&= \mathrm{E}_x^{n\,2} + \mathrm{E}_y^{n\,2} + \mathrm{E}_z^{n\,2} \\
&= \left(\mathbf{E}_x^n\mathbf{V}\right)^{\mathrm{T}}\mathbf{E}_x^n\mathbf{V} + \left(\mathbf{E}_y^n\mathbf{V}\right)^{\mathrm{T}}\mathbf{E}_y^n\mathbf{V} + \left(\mathbf{E}_z^n\mathbf{V}\right)^{\mathrm{T}}\mathbf{E}_z^n\mathbf{V} \\
&= \mathbf{V}^{\mathrm{T}}\mathbf{E}_x^{n\mathrm{T}}\mathbf{E}_x^n\mathbf{V} + \mathbf{V}^{\mathrm{T}}\mathbf{E}_y^{n\mathrm{T}}\mathbf{E}_y^n\mathbf{V} + \mathbf{V}^{\mathrm{T}}\mathbf{E}_z^{n\mathrm{T}}\mathbf{E}_z^n\mathbf{V} \\
&= \mathbf{V}^{\mathrm{T}}\left(\mathbf{E}_x^{n\mathrm{T}}\mathbf{E}_x^n + \mathbf{E}_y^{n\mathrm{T}}\mathbf{E}_y^n + \mathbf{E}_z^{n\mathrm{T}}\mathbf{E}_z^n\right)\mathbf{V}
\end{aligned}
\tag{3.26}
$$

The original microwave heating power P_{mw}^n can be rewritten using equation 3.26 as

$$
\begin{aligned}
P_{\mathrm{mw}}^n &= \frac{1}{2}l^3\,\sigma_{\mathrm{e}}(T)\cdot\left\|\vec{\mathbf{E}}^n\right\|^2 \\
&= \frac{1}{2}l^3\,\sigma_{\mathrm{e}}(T)\cdot\mathbf{V}^{\mathrm{T}}\left(\mathbf{E}_x^{n\mathrm{T}}\mathbf{E}_x^n + \mathbf{E}_y^{n\mathrm{T}}\mathbf{E}_y^n + \mathbf{E}_z^{n\mathrm{T}}\mathbf{E}_z^n\right)\mathbf{V} \\
&= \mathbf{V}^{\mathrm{T}}\left[\mathbf{\Phi}_{\mathrm{c}}^n(T)\right]\mathbf{V},
\end{aligned}
\tag{3.27}
$$

where the matrix $[\mathbf{\Phi}_{\mathrm{c}}^n(T)]$ is defined by

$$
\left[\mathbf{\Phi}_{\mathrm{c}}^n(T)\right] := \frac{1}{2}l^3\,\sigma_{\mathrm{e}}(T)\cdot\left(\mathbf{E}_x^{n\mathrm{T}}\mathbf{E}_x^n + \mathbf{E}_y^{n\mathrm{T}}\mathbf{E}_y^n + \mathbf{E}_z^{n\mathrm{T}}\mathbf{E}_z^n\right). \tag{3.28}
$$

The matrix $[\mathbf{\Phi}_{\mathrm{c}}^n(T)]$ is a $M \times M$ symmetric square matrix, which is the effective heating matrix at this location n, and $\mathbf{V}$ is the control input vector that has to be calculated by a controller. In contrast to the case of scalar superposition, the vector superposition rule is much more complicated due to its high computation complexity. However, it is accurate and effective to describe the power superposition of microwave heating systems with multiple feeding sources, as shown by the simulation results in [BPJD99] and [THN01].

3.2.3. Formulation and Discretization

In this section, above approximations 3.15, 3.18 and 3.27 will be further formulated and the resulted equations will be discretized, to make sure they can be directly used in the following controller design. For each location n, the original heat transfer equation 3.12 is converted into two different forms, regarding different approximations. For the case of scalar superposition, the corresponding linear heat transfer equation is shown as

$$\rho c_\mathrm{p} \frac{\mathrm{d}T^n}{\mathrm{d}t} \cdot l^3 = -\tilde{h}^n(T^n)\,(T^n - T_\mathrm{a}) \cdot l^2 + \mathbf{P}^n_\mathrm{mw}\mathbf{U}, \tag{3.29}$$

where T^n is the measured temperature at the n-th location, and $-\tilde{h}^n(T^n)$ is the power dissipation coefficient. Defining the new variable $Y^n = T^n - T_\mathrm{a}$, the above equation is rewritten as

$$\rho c_\mathrm{p} \frac{\mathrm{d}Y^n}{\mathrm{d}t} \cdot l^3 = -\hat{h}^n(Y^n) \cdot Y^n \cdot l^2 + \mathbf{P}^n_\mathrm{mw}\mathbf{U}, \tag{3.30}$$

with

$$\hat{h}^n(Y^n) = h + \varrho\sigma'\left[(Y^n + T_\mathrm{a})^2 + T_\mathrm{a}^2\right] \cdot (Y^n + 2T_\mathrm{a}) \tag{3.31}$$

In microwave heating applications, the temperatures of the load are always changing along time. All temperature dependent variables in equation 3.30, as well as the heating power vector $\mathbf{P}^n_\mathrm{mw}$ and the control input vector $\mathbf{U}$, can be written as time varying variables. Therefore equation 3.30 can be formulated into a continuous-time equation, such as

$$\frac{\mathrm{d}Y^n(t)}{\mathrm{d}t} = \frac{-\hat{h}^n(t)}{\rho c_\mathrm{p}\, l} \cdot Y^n(t) + \frac{1}{\rho c_\mathrm{p} l^3} \cdot \mathbf{P}^n_\mathrm{mw}(t)\mathbf{U}(t). \tag{3.32}$$

From another point of view, in the above equation all these time-varying parameters, e.g. $Y^n(t)$, represent not functions depending on time, but rather values of vectors or matrices at the time t.

Defining

$$\begin{aligned} A^n_\mathrm{c}(t) &:= \frac{-\hat{h}^n(t)}{\rho c_\mathrm{p}\, l}, \\ \mathbf{B}^n_\mathrm{c}(t) &:= \frac{1}{\rho c_\mathrm{p} l^3} \cdot \mathbf{P}^n_\mathrm{mw}(t) \\ &= \left[\, B^{n,1}_\mathrm{c}(t),\, B^{n,2}_\mathrm{c}(t), \ldots, B^{n,M}_\mathrm{c}(t) \,\right], \end{aligned} \tag{3.33}$$

then equation 3.32 is further transferred into

$$\frac{\mathrm{d}Y^n}{\mathrm{d}t} = A_\mathrm{c}^n(t) \cdot Y^n + \mathbf{B}_\mathrm{c}^n(t)\mathbf{U}(t). \tag{3.34}$$

Based on this single output model, a continuous-time MIMO model is built as

$$\frac{\mathrm{d}\mathbf{Y}}{\mathrm{d}t} = [\mathbf{A}_\mathrm{c}(t)]\,\mathbf{Y} + [\mathbf{B}_\mathrm{c}(t)]\,\mathbf{U}(t), \tag{3.35}$$

where $\mathbf{Y}$ is the output vector defined as

$$\mathbf{Y} = \left[Y^1, Y^2, \ldots, Y^N\right]^\mathrm{T}, \tag{3.36}$$

and the state matrix and the output matrix are defined as

$$\begin{aligned}
[\mathbf{A}_\mathrm{c}(t)] &:= \begin{bmatrix} A_\mathrm{c}^1(t) & 0 & \cdots & 0 \\ 0 & A_\mathrm{c}^2(t) & \cdots & 0 \\ \vdots & \vdots & \ddots & \vdots \\ 0 & 0 & \cdots & A_\mathrm{c}^N(t) \end{bmatrix}, \\
[\mathbf{B}_\mathrm{c}(t)] &:= \begin{bmatrix} B_\mathrm{c}^{1,1}(t) & B_\mathrm{c}^{1,2}(t) & \cdots & B_\mathrm{c}^{1,M}(t) \\ B_\mathrm{c}^{2,1}(t) & B_\mathrm{c}^{2,2}(t) & \cdots & B_\mathrm{c}^{2,M}(t) \\ \vdots & \vdots & \ddots & \vdots \\ B_\mathrm{c}^{N,1}(t) & B_\mathrm{c}^{N,2}(t) & \cdots & B_\mathrm{c}^{N,M}(t) \end{bmatrix}.
\end{aligned} \tag{3.37}$$

Before above models can be used in real controller design, the last step is discretization, because in practice the controller is always operated in discrete-time. Different discretization methods exist and in this dissertation the Euler method [Lju98] is applied. When the sampling time Δt is a sufficiently small constant (1.5 s or even less), the temperature changing rate on the left-hand side of equation 3.35 can be approximated by

$$\frac{\mathrm{d}\mathbf{Y}}{\mathrm{d}t} \approx \frac{1}{\Delta t}\Big(\mathbf{Y}(k+1) - \mathbf{Y}(k)\Big),$$

where k indicates the discrete time interval.

The final linear discrete-time MIMO heat transfer model is

$$\begin{aligned}\mathbf{Y}(k+1) &= \Big([\mathbf{I}_N] + \Delta t\ [\mathbf{A}_\mathrm{c}(k)] \Big)\mathbf{Y}(k) + \Delta t\ [\mathbf{B}_\mathrm{c}(k)]\ \mathbf{U}(k), \\ &= [\mathbf{A}(k)]\ \mathbf{Y}(k) + [\mathbf{B}(k)]\ \mathbf{U}(k)\end{aligned} \tag{3.38}$$

with

$$[\mathbf{A}(k)] := [\mathbf{I}_N] + \Delta t\ [\mathbf{A}_\mathrm{c}(k)] = \begin{bmatrix} A^1(k) & 0 & \cdots & 0 \\ 0 & A^2(k) & \cdots & 0 \\ \vdots & \vdots & \ddots & \vdots \\ 0 & 0 & \cdots & A^N(k) \end{bmatrix},$$

$$[\mathbf{B}(k)] := \Delta t \cdot [\mathbf{B}_\mathrm{c}(k)] = \begin{bmatrix} B^{1,1}(k) & B^{1,2}(k) & \cdots & B^{1,M}(k) \\ B^{2,1}(k) & B^{2,2}(k) & \cdots & B^{2,M}(k) \\ \vdots & \vdots & \ddots & \vdots \\ B^{N,1}(k) & B^{N,2}(k) & \cdots & B^{N,M}(k) \end{bmatrix}. \tag{3.39}$$

and

$$A^n(k) = 1 + \Delta t \cdot A^n_\mathrm{c}(k), \qquad B^{n,m}(k) = \Delta t \cdot B^{n,m}_\mathrm{c}(k). \tag{3.40}$$

The matrix $[\mathbf{I}_N]$ is an identity matrix with the dimension $N \times N$. The equation 3.38 indicates that the future temperature vector $\mathbf{Y}(k+1)$ at time $k+1$ can be determined by the current temperature vector $\mathbf{Y}(k)$ and the control input vector $\mathbf{U}(k)$ at time k. Therefore the current temperature vector $\mathbf{Y}(k)$ (equation 3.38) is written as

$$\mathbf{Y}(k) = [\mathbf{A}(k-1)]\ \mathbf{Y}(k-1) + [\mathbf{B}(k-1)]\ \mathbf{U}(k-1), \tag{3.41}$$

which will be used later (section 3.2.4) for the linear system identification derivation.

Following the similar derivation process, a nonlinear discrete-time MIMO model can also be constructed such as

$$\mathbf{Y}(k+1) = [\mathbf{A}(k)]\ \mathbf{Y}(k) + \mathbf{\Psi}(k), \tag{3.42}$$

where $\mathbf{Y}(k)$ and $\mathbf{A}(k)$ are the same as defined in the linear model (equation 3.41). The vector $\boldsymbol{\Psi}$ is a augmented column vector which is defined as

$$\begin{aligned}\boldsymbol{\Psi}(k) &:= \left[\Psi^1(k), \Psi^2(k), \ldots, \Psi^N(k)\right]^{\mathrm{T}},\\ \Psi^n(k) &:= \frac{\Delta t}{\rho c_{\mathrm{p}} l^3} \cdot \mathbf{V}^{\mathrm{T}}(k)\left[\boldsymbol{\Phi}_{\mathrm{c}}^n(k)\right]\mathbf{V}(k), \quad 1 \leq n \leq N.\end{aligned} \tag{3.43}$$

In order to have a more clear notation, the discrete-time effective heating matrix $[\boldsymbol{\Phi}^n(k)]$ is defined as

$$[\boldsymbol{\Phi}^n(k)] := \frac{\Delta t}{\rho c_{\mathrm{p}} l^3} \cdot [\boldsymbol{\Phi}_{\mathrm{c}}^n(k)], \tag{3.44}$$

which leads to

$$\Psi^n(k) = \mathbf{V}^{\mathrm{T}}(k)\left[\boldsymbol{\Phi}^n(k)\right]\mathbf{V}(k). \tag{3.45}$$

Similar to equation 3.41, the current temperature vector (equation 3.42) is represented as

$$\mathbf{Y}(k) = [\mathbf{A}(k-1)]\ \mathbf{Y}(k-1) + \boldsymbol{\Psi}(k-1), \tag{3.46}$$

which will be used for the nonlinear system identification.

The grey-box modeling approach starts from the very basic heat transfer model, combining processes represented by equation 3.12. It neglects minor influences (thermal conduction) and approximates complex heating terms with computable expressions. In the end, two discrete-time models (equations 3.41 and 3.46) are generated. If all unknown matrices in the model, e.g. the matrix $[\mathbf{A}(k)]$, are accurately estimated, the control input vector $\mathbf{U}(k)$ or $\mathbf{V}(k)$ can be calculated to control the system for achieving the desired target.

3.2.4. Online System Identification

All unknown parameters and matrices in a system model can be estimated using different system identification algorithms. With the estimated parameters, the future behaviors of the system can be predicted and the system can be controlled accordingly. In order to get a good control performance for the temperature control system of HEPHAISTOS, the system identification algorithm used in our case needs to fulfill the following requirements:

- Implementation in a recursive or online form.
- Ability of dealing with high-dimensional systems.
- Ability of tracking time-varying behaviors.
- Guaranteed convergence and fast converging speed.

According to these four requirements, the exponentially weighted recursive least squares (RLS) [Lju98], recursive Kalman filter (RKF) [Lju98] and extended Kalman filter (EKF) [Lju98] [WVDM00] algorithms were selected as the system identification algorithms, with respect to the two different models (equations 3.41 and 3.46).

Linear Recursive System Identification

The task of the linear recursive system identification is to estimate the value of matrices $[\mathbf{A}(k)]$ and $[\mathbf{B}(k)]$ in a online (recursive) form. In the linear model (equation 3.41), both matrices are assumed to be time-varying. Since there is no prior knowledge indicating the varying trend of $[\mathbf{A}(k)]$ or $[\mathbf{B}(k)]$, the best choice is to assume both matrices fulfill the following equations [ZL03]

$$\begin{aligned} [\mathbf{A}(k)] &= [\mathbf{A}(k-1)] + [\varepsilon_1(k-1)]\,, \\ [\mathbf{B}(k)] &= [\mathbf{B}(k-1)] + [\varepsilon_2(k-1)]\,. \end{aligned} \tag{3.47}$$

Both $[\varepsilon_1(k-1)]$ and $[\varepsilon_2(k-1)]$ are zero-mean white noise added at time $k-1$, and their covariance matrices are donated as $[\mathbf{\Omega}_1]$ and $[\mathbf{\Omega}_2]$ respectively. At the current time k, the matrices $[\mathbf{A}(k-1)]$ and $[\mathbf{B}(k-1)]$ can be estimated using the current temperature vector $\mathbf{Y}(k)$ and historical data sets $\{\mathbf{U}(k-1),\ \mathbf{Y}(k-1),\ \mathbf{U}(k-2),\ \mathbf{Y}(k-2), \dots\}$. Then the future matrices $[\mathbf{A}(k)]$ and $[\mathbf{B}(k)]$ can be predicted using these estimations.

The system identification process applied here is done in a multiple-input single output (MISO) way. In the beginning, the linear MIMO model (equation 3.41) has to be transformed into N different MISO ARX models (AutoRegressive model with eXogenous inputs) [BJR13], such as

$$\begin{aligned} Y^n(k) &= A^n(k-1)Y^n(k-1) + \mathbf{B}^n(k-1)\mathbf{U}(k-1) \\ &= \boldsymbol{\theta}^n(k-1)\,\boldsymbol{\Pi}^n(k-1), \end{aligned} \tag{3.48}$$

$$\begin{aligned} \boldsymbol{\theta}^n(k-1) &:= \Big[A^n(k-1),\, \mathbf{B}^n(k-1)\Big], \\ &= \Big[A^n(k-1),\, B^{n,1}(k-1),\, \ldots, B^{n,M}(k-1)\Big], \end{aligned} \tag{3.49}$$

$$\begin{aligned} \boldsymbol{\Pi}^n(k-1) &:= \Big[Y^n(k-1),\, \mathbf{U}^{\mathrm{T}}(k-1)\Big]^{\mathrm{T}}, \\ &= \Big[Y^n(k-1),\, u_1(k-1),\, \ldots, u_M(k-1)\Big]^{\mathrm{T}}, \end{aligned} \tag{3.50}$$

where $Y^n(k)$ is the true temperature value at time k, $\boldsymbol{\theta}^n(k-1)$ is the augmented coefficient vector for the output $Y^n(k)$ with the dimension $1 \times (1+M)$, and $\boldsymbol{\Pi}^n(k-1)$ is the augmented data vector with the dimension $(1+M) \times 1$. According to the assumptions made in equations 3.47, the coefficient vector $\boldsymbol{\theta}^n(k)$ can also be represented as a varying variable, such as

$$\begin{aligned} \boldsymbol{\theta}^n(k) &= \boldsymbol{\theta}^n(k-1) + \varepsilon(k-1), \\ \boldsymbol{\theta}^n(k-1) &= \boldsymbol{\theta}^n(k-2) + \varepsilon(k-2), \end{aligned} \tag{3.51}$$

where ε is a zero-mean white noise with the covariance matrix $[\boldsymbol{\Omega}]$. The first equation indicates that the future augmented coefficient vector $\boldsymbol{\theta}^n(k)$ can be predicted based on the current $\boldsymbol{\theta}^n(k-1)$. The second equations means that the current $\boldsymbol{\theta}^n(k-1)$ can be estimated using the former vector $\boldsymbol{\theta}^n(k-2)$. The definitions of estimation and prediction as well as other parameters used in the linear system identification can be found in table 3.1.

In equation 3.48, both $Y^n(k)$ and $\boldsymbol{\Pi}^n(k-1)$ are known at time k. The unknown augmented vector $\boldsymbol{\theta}^n(k-1)$ has to be estimated, from which both $A^n(k-1)$ and $\mathbf{B}(k-1)$ can be obtained simultaneously. In practice, the real measured temperature value $Y_{\mathrm{r}}^n(k)$ is written as

$$Y_{\mathrm{r}}^n(k) = Y^n(k) + \varsigma(k) = \boldsymbol{\theta}^n(k-1)\,\boldsymbol{\Pi}^n(k-1) + \varsigma(k), \tag{3.52}$$

Definition	Interpretation
$\boldsymbol{\theta}_{\mathrm{e}}^{n}(k-1) = E\big[\boldsymbol{\theta}^{n}(k-1) \big\vert\, Y_{\mathrm{r}}^{n}(k)\big]$	Estimation: expectation of $\boldsymbol{\theta}^{n}(k-1)$ based on the current measured temperature $Y_{\mathrm{r}}^{n}(k)$
$\boldsymbol{\theta}_{\mathrm{p}}^{n}(k-1) = E\big[\boldsymbol{\theta}^{n}(k-1) \big\vert\, Y_{\mathrm{r}}^{n}(k-1)\big]$	Prediction: expectation of $\boldsymbol{\theta}^{n}(k-1)$ based on the former measured temperature $Y_{\mathrm{r}}^{n}(k-1)$
$\mathbf{e}_{\mathrm{e}}^{n}(k-1) = \boldsymbol{\theta}^{n}(k-1) - \boldsymbol{\theta}_{\mathrm{e}}^{n}(k-1)$	Estimation error between $\boldsymbol{\theta}^{n}(k-1)$ and $\boldsymbol{\theta}_{\mathrm{e}}^{n}(k-1)$
$\mathbf{e}_{\mathrm{p}}^{n}(k-1) = \boldsymbol{\theta}^{n}(k-1) - \boldsymbol{\theta}_{\mathrm{p}}^{n}(k-1)$	Prediction error between $\boldsymbol{\theta}^{n}(k-1)$ and $\boldsymbol{\theta}_{\mathrm{p}}^{n}(k)$
$[\mathbf{P}_{\mathrm{e}}^{n}(k-1)] = E\left[\mathbf{e}_{\mathrm{e}}^{n}(k-1)\big(\mathbf{e}_{\mathrm{e}}^{n}(k-1)\big)^{\mathrm{T}}\right]$	Covariance matrix of the estimation error $\mathbf{e}_{\mathrm{e}}^{n}(k-1)$
$[\mathbf{P}_{\mathrm{p}}^{n}(k-1)] = E\left[\mathbf{e}_{\mathrm{p}}^{n}(k-1)\big(\mathbf{e}_{\mathrm{p}}^{n}(k-1)\big)^{\mathrm{T}}\right]$	Covariance matrix of the prediction error $\mathbf{e}_{\mathrm{p}}^{n}(k-1)$

Table 3.1. Parameters and vectors used in the linear system identification

where $\varsigma(k)$ is the measurement error at time k. It is also assumed that the measurement error has zero mean value and a covariance of σ^2. The currently available estimation of the temperature value is

$$Y_{\mathrm{e}}^{n}(k) = \boldsymbol{\theta}_{\mathrm{e}}^{n}(k-2)\,\boldsymbol{\Pi}^{n}(k-1), \tag{3.53}$$

where $\boldsymbol{\theta}_{\mathrm{e}}^{n}(k-2)$ is the former estimated coefficient vector. Following the procedures in [KV86], the new estimation $\boldsymbol{\theta}_{\mathrm{e}}^{n}(k-1)$ can be obtained using the following equation

$$\boldsymbol{\theta}_{\mathrm{e}}^{n}(k-1) = \boldsymbol{\theta}_{\mathrm{e}}^{n}(k-2) + \mathbf{K}^{n}(k-1)\Big[Y_{\mathrm{r}}^{n}(k) - Y_{\mathrm{e}}^{n}(k)\Big], \tag{3.54}$$

where $\mathbf{K}^n(k-1)$ is defined as the estimation gain vector at time k with the dimension $1 \times (1+M)$. Equation 3.54 indicates that the new estimation $\boldsymbol{\theta}_{\mathrm{e}}^n(k-1)$ can be obtained using the former estimation $\boldsymbol{\theta}_{\mathrm{e}}^n(k-2)$ plus a correction term. This correction term is proportional to the gain vector $\mathbf{K}^n(k-1)$, as well as the difference between $Y_{\mathrm{r}}^n(k)$ and $Y_{\mathrm{e}}^n(k)$. A brief illustration of the linear system identification is represented by figure 3.3.

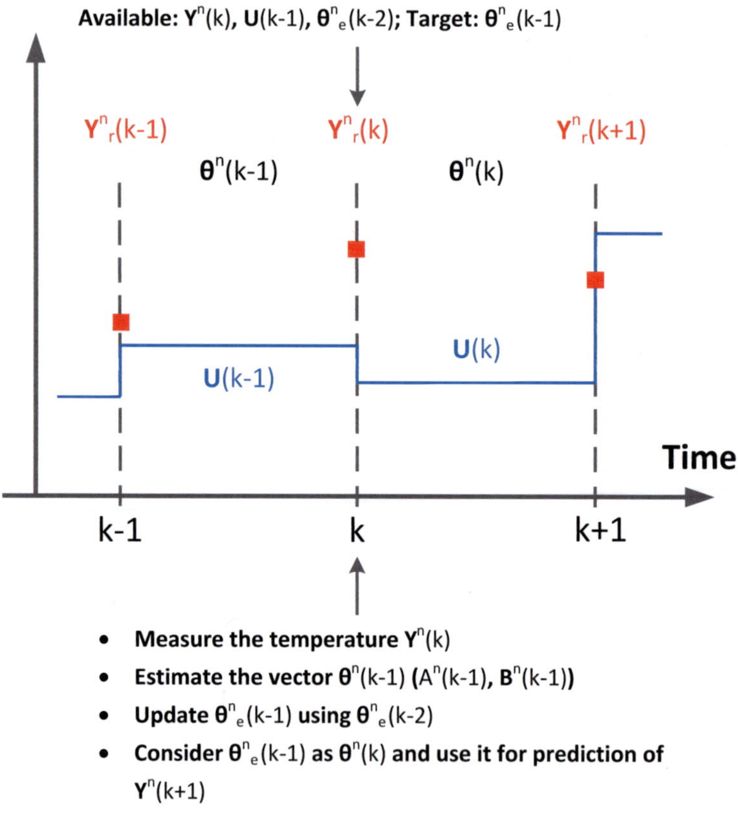

Figure 3.3. Illustration of the linear system identification process.

Two linear recursive system identification algorithms have been analyzed and tested in our experiments.

- Exponentially Weighted Recursive Least Squares (RLS)

  The exponentially weighted recursive least squares (RLS) algorithm is one of the most used algorithms in adaptive filtering and system identification [Li08], for tracking time-varying parameters. Given all observations $\big(Y^n(i), \mathbf{\Pi}^n(i)\big)$ from the beginning $(i = 1)$ to the current time $(i = k)$, the cost function $J_{\mathrm{RLS}}(k)$ of the exponentially weighted RLS is defined as [Lju98]

$$J_{\mathrm{RLS}}(k) = \frac{1}{k-1} \sum_{i=2}^{k} \lambda^{k-i} \Big(Y_{\mathrm{r}}^n(i) - \boldsymbol{\theta}_{\mathrm{e}}^n(i-1)\, \mathbf{\Pi}^n(i-1) \Big)^2, \quad (3.55)$$

  where λ is a forgetting factor with $0 < \lambda \leq 1$.

  A cost function is the function defined to be minimized. The minimization of the cost function leads to the optimal coefficient vector $\boldsymbol{\theta}_{\mathrm{e}}^n$, such as

$$\boldsymbol{\theta}_{\mathrm{e}}^n(k-1) = \arg_{\boldsymbol{\theta}} \min J_{\mathrm{RLS}}(k). \quad (3.56)$$

  The involvement of λ indicates that the above cost function assigns more credits to recent data than old data, endowing the exponentially weighted RLS the ability of tracking time-varying systems. The smaller λ is, the faster it forgets old data.

  The detailed derivation process of the estimation $\boldsymbol{\theta}_{\mathrm{e}}^n(k)$ can be found in [WP97]. The final update equations of the exponentially weighted RLS with $k \geq 2$ are given as [Pol03]

$$\begin{aligned}
\mathbf{K}^n(k-1) &= [\mathbf{P}_{\mathrm{e}}^n(k-2)]\, \mathbf{\Pi}(k-1) \big[\lambda\sigma^2 \\
&\qquad + \mathbf{\Pi}^{\mathrm{T}}(k-1)\, [\mathbf{P}_{\mathrm{e}}^n(k-2)]\, \mathbf{\Pi}(k-1) \big]^{-1}, \\
\boldsymbol{\theta}_{\mathrm{e}}^n(k-1) &= \boldsymbol{\theta}_{\mathrm{e}}^n(k-2) + \mathbf{K}^n(k-1) \Big[Y_{\mathrm{r}}^n(k) - Y_{\mathrm{e}}^n(k) \Big], \\
[\mathbf{P}_{\mathrm{e}}^n(k-1)] &= \frac{1}{\lambda} \big[1 - \mathbf{K}^n(k-1)\mathbf{\Pi}(k-1) \big]\, [\mathbf{P}_{\mathrm{e}}^n(k-2)]\,.
\end{aligned} \quad (3.57)$$

  In practice, the vector $\boldsymbol{\theta}_{\mathrm{e}}^n(0)$ is randomly initialized. and the covariance matrix $[\mathbf{P}_{\mathrm{e}}^n]$ is initialized as $[\mathbf{P}_{\mathrm{e}}^n(0)] = r\,[\mathbf{I}_{1+M}]$, where r is a real number [Lju98]. The value of r denotes the level of the initial estimation error. For example, a large r indicates a large estimation error and then the estimation $\boldsymbol{\theta}_{\mathrm{e}}^{n}$ will jump away from the current state quickly. In this dissertation $r = 100$ is used.

- Recursive Kalman filter (RKF)

  Kalman filter [Rib04] is another powerful algorithm that has been widely implemented in adaptive filtering, adaptive control and system identification areas. In RKF the system is assumed the same as in the RLS algorithm as

$$Y_{\mathrm{r}}^{n}(k) = \boldsymbol{\theta}^{n}(k-1)\,\boldsymbol{\Pi}^{n}(k-1) + \varsigma(k), \tag{3.58}$$
$$\boldsymbol{\theta}^{n}(k-1) = \boldsymbol{\theta}^{n}(k-2) + \varepsilon(k-2), \tag{3.59}$$

  where $\varsigma(k)$ is the measurement error as in equation 3.52 and $\varepsilon(k-2)$ is the white noise as in equation 3.51. This assumption also provides RKF the ability to track time-varying parameters.

  The complete update process of RKF can be divided into two parts. The first part is the prediction of the unknown vectors at the current time k, based on the data obtained at the former time step $k-1$, as given in [Rib04] with $k \geq 2$

$$\begin{aligned} \boldsymbol{\theta}_{\mathrm{p}}^{n}(k-1) &= \boldsymbol{\theta}_{\mathrm{e}}^{n}(k-2), \\ \left[\mathbf{P}_{\mathrm{p}}^{n}(k-1)\right] &= \left[\mathbf{P}_{\mathrm{e}}^{n}(k-2)\right] + [\boldsymbol{\Omega}]\,. \end{aligned} \tag{3.60}$$

  The first equation in 3.60 indicates that the predicted coefficient vector $\boldsymbol{\theta}_{\mathrm{p}}(k-1)$ based on the old data is the equivalent to the previous estimation $\boldsymbol{\theta}_{\mathrm{e}}(k-2)$. The influence of the unknown noise variable $\varepsilon(k-1)$ is reflected in the second equation, where the covariance matrix of prediction error $\mathbf{P}_{\mathrm{p}}(k)$ equals to the covariance matrix of the previous estimation error $\mathbf{P}_{\mathrm{e}}(k-1)$ plus the covariance matrix of the noise $\boldsymbol{\Omega}$.

  The second part is called estimation, which is to estimate the current variables using the currently measured temperatures and former predictions, such as given in [Rib04] with $k \geq 2$

$$\begin{aligned}
\mathbf{K}^n(k-1) = {} & \left[\mathbf{P}^n_{\mathrm{p}}(k-1)\right]\mathbf{\Pi}(k-1)\big[\sigma^2 \\
& + \mathbf{\Pi}^{\mathrm{T}}(k-1)\left[\mathbf{P}^n_{\mathrm{p}}(k-1)\right]\mathbf{\Pi}(k-1)\big]^{-1}, \\
\boldsymbol{\theta}^n_{\mathrm{e}}(k-1) = {} & \boldsymbol{\theta}^n_{\mathrm{p}}(k-1) + \mathbf{K}^n(k-1)\Big[Y^n_{\mathrm{r}}(k) - \boldsymbol{\theta}^n_{\mathrm{p}}(k-1)\mathbf{\Pi}(k-1)\Big], \\
\left[\mathbf{P}^n_{\mathrm{e}}(k-1)\right] = {} & \left[1 - \mathbf{K}^n(k-1)\mathbf{\Pi}(k-1)\right]\left[\mathbf{P}^n_{\mathrm{p}}(k-1)\right].
\end{aligned} \tag{3.61}$$

The vectors updated from the prediction part (equations 3.60) can be directly used in the estimation part (equations 3.61). After the estimation, the estimated coefficient vector $\boldsymbol{\theta}^n_{\mathrm{e}}(k-1)$ can be employed to predict future temperature value $Y^n(k+1)$ because of $\boldsymbol{\theta}^n_{\mathrm{p}}(k) = \boldsymbol{\theta}^n_{\mathrm{e}}(k-1)$. Both $[\boldsymbol{\Omega}]$ and σ^2 can be either predefined by the user or estimated online using other approaches [ÅJPJ08]. When the covariances of the noise signals are perfectly known, RKF is guaranteed to provide the optimal estimation [Rib04]. But in practice, this condition is not always fulfilled. Normally the covariances $[\boldsymbol{\Omega}]$ and σ^2 are used as the tuning elements of RKF to adjust the estimation properties. Detailed derivation process of RKF refers to [WVDM00] or [Rib04].

It should be noted that, if λ is set to be 1 and the matrix $[\boldsymbol{\Omega}]$ is set to be a zero matrix, the two results (equations 3.55 and 3.61) would be exactly the same. In other words, the RLS solution coincidis es with the RKF solution when the coefficient vector $\boldsymbol{\theta}^n(k)$ is deterministic and the cost function $J_{\mathrm{RLS}}(k)$ (equation 3.55) is non-weighted.

As mentioned in the beginning of this section, the solutions of RLS and RKF given by equations 3.55 and 3.61 are expressed in a MISO form. For a MIMO model that contains N different measured temperatures, the update process has to be repeated for N times per estimation period. An alternative to the MISO-form system identification is to replace equation 3.48 by the MIMO equation 3.41, and directly apply RLS or RKF to the MIMO ARX model. In this case, only one update process is needed per estimation period. But the problem is that the state matrix $[\mathbf{A}(k-1)]$ resulted from the MIMO update approach is not diagonal, which is against equation 3.39. Correspondingly the estimation result of the MIMO-form system identification is significantly different from the result of the MISO-form. The performance of both RLS and RKF is compared in real experiments, which will be presented in chapter 4.

Nonlinear Recursive System Identification

The task of nonlinear recursive system identification is to estimate the matrices $[\mathbf{A}^n(k-1)]$ and $[\mathbf{\Phi}^n(k-1)]$ in equation 3.46. Compared with the linear case, the nonlinear system identification process is more complicated because of the nonlinear heating term $\mathbf{\Psi}$. There is no way to combine the parameter $A^n(k-1)$ and the effective heating matrix $[\mathbf{\Phi}^n(k-1)]$ in one matrix (as in the vector $\boldsymbol{\theta}^n(k)$), and estimate them at the same time. Instead, in the nonlinear system identification, a two-step identification procedure has been developed, to estimate them separately at different time using the extended Kalman filter (EKF) method. This estimation is also applied on the MISO-form model, which is similar as in the linear case.

- Step 1: Estimate the first part $[\mathbf{A}(k-1)]$

  This step is done during the cooling part or whenever no power is injected from the microwave sources ($v_m(k) = 0$ for all $1 \leq m \leq M$). Since the input vector is zero, the nonlinear MIMO system model (equation 3.46) can be decomposed into N MISO systems, such as

  $$Y^n(k) = A^n(k-1)\,Y^n(k-1), \quad 1 \leq n \leq N. \tag{3.62}$$

  The parameter $A^n(k-1)$ can be estimated directly as

  $$A^n(k-1) = Y^n(k)/Y^n(k-1), \quad 1 \leq n \leq N. \tag{3.63}$$

- Step 2: Estimate the effective heating matrix $[\mathbf{\Phi}^n(k-1)]$

  When the input vector $\mathbf{V}$ is not zero, the state parameter $A^n(k-1)$ is assumed to be constant like $A^n(k-1) = A^n_{\text{ct}}$, where A^n_{ct} is the last updated value of A^n. This assumption is valid in practice because the parameter $A^n(k-1)$ is mainly determined by the convection heat transfer coefficient h (see equation 3.15 and the comparison in equation 3.14) which is a constant. The varying part of $A^n(k-1)$

caused by different thermal radiation effects can be omitted, especially when the temperature is in a stable range. Then the n-th MISO systems can be written as

$$\begin{aligned} Y^n(k) &= A^n_{\text{ct}} \cdot Y^n(k-1) + \Psi^n(k-1), \\ \Psi^n(k-1) &= \mathbf{V}^{\text{T}}(k-1)\,[\mathbf{\Phi}^n(k-1)]\,\mathbf{V}(k-1). \end{aligned} \tag{3.64}$$

At any time k, the value of $\Psi^n(k-1)$ can be calculated by

$$\Psi^n(k-1) = Y^n(k) - A^n_{\text{ct}} \cdot Y^n(k-1), \tag{3.65}$$

and then the effective heating matrix $[\mathbf{\Phi}^n(k-1)]$ can be estimated according to the value of $\Psi^n(k-1)$.

As in the linear RKF (equation 3.59), in EKF it is also assumed that the system fulfills the following equations

$$\begin{aligned} \Psi^n_{\text{r}}(k-1) &= \mathbf{V}^{\text{T}}(k-1)\,[\mathbf{\Phi}^n(k-1)]\,\mathbf{V}(k-1) + \varsigma(k-1), \\ [\mathbf{\Phi}^n(k-1)] &= [\mathbf{\Phi}^n(k-2)] + [\varepsilon(k-1)], \end{aligned} \tag{3.66}$$

where $\Psi^n_{\text{r}}(k-1)$ is the real calculated value of $\Psi^n(k-1)$, $\varsigma(k-1)$ is a zero-mean measurement or calculation error with the covariance σ^2 and $[\varepsilon(k-1)]$ is a zero-mean white noise with the covariance matrix $[\mathbf{\Omega}]$. The definitions of parameters and vectors used in EKF is shown by table 3.2.

The detailed derivation process is enclosed in the appendix A.1. The update rule of EKF is given as following.

Prediction part with $k \geq 2$:

$$\begin{aligned} \left[\mathbf{\Phi}^n_{\text{p}}(k-1)\right] &= [\mathbf{\Phi}^n_{\text{e}}(k-2)], \\ \left[\mathbf{P}^n_{\text{p}}(k-1)\right] &= [\mathbf{P}^n_{\text{e}}(k-2)] + [\mathbf{\Omega}]. \end{aligned} \tag{3.67}$$

Estimation part with $k \geq 2$:

$$
\begin{aligned}
\left[\mathbf{K}^n(k-1)\right] = &\left[\mathbf{P}_{\mathrm{p}}^n(k-1)\right]\left[\mathbf{J}^n(k-1)\right]^{\mathrm{T}}\Big(\left[\mathbf{J}^n(k-1)\right]\left[\mathbf{P}_{\mathrm{p}}^n(k-1)\right] \\
&\cdot\left[\mathbf{J}^n(k-1)\right]^{\mathrm{T}}+\sigma^2\left[\mathbf{I}_M\right]\Big)^{-1}, \\
\left[\mathbf{\Phi}_{\mathrm{e}}^n(k-1)\right] = &\left[\mathbf{\Phi}_{\mathrm{p}}^n(k-1)\right]+\left[\mathbf{K}^n(k-1)\right]\Big(\Psi_{\mathrm{r}}^n(k-1) \\
&-\mathbf{V}^{\mathrm{T}}(k-1)\left[\mathbf{P}_{\mathrm{p}}^n(k-1)\right]\mathbf{V}(k-1)\Big), \\
\left[\mathbf{P}_{\mathrm{e}}^n(k-1)\right] = &\Big(\left[\mathbf{I}_M\right]-\left[\mathbf{K}^n(k-1)\right]\left[\mathbf{J}^n(k-1)\right]\Big)\left[\mathbf{P}_{\mathrm{p}}^n(k-1)\right],
\end{aligned}
\tag{3.68}
$$

with

$$
\left[\mathbf{J}^n(k-1)\right] := \mathbf{V}(k-1)\mathbf{V}^{\mathrm{T}}(k-1).
$$

According to equation 3.67, at each time k the estimated effective matrix $\mathbf{\Phi}_{\mathrm{e}}^n(k-1)$ is expected to be equivalent to $\mathbf{\Phi}^n(k)$, and then used for the prediction of future temperatures.

3.3. Black-box Modeling

Although grey-box modeling is simple to implement and control, it has several constraints, such as the fixed type of setup (figure 3.1) and restrictions such as the assumptions 1 and 2. Compared with the grey-box modeling, black-box modeling is much more flexible, without any restrictions in the setup type. In order to enhance the adaptivity of the temperature control system and improve the model accuracy, the black-box modeling approach is also applied.

The scope of black-box structures is diverse, covering methods from neural networks, radial basis networks, wavelet networks, to fuzzy logic models. These different methods are used not only for modeling, but also for pattern classification, controller design, self-learning systems, and artificial intelligence systems. The principle used in most black-box methods is a two-layer mapping. The first mapping

Definition	Interpretation
$[\mathbf{\Phi}_{\mathrm{e}}^{n}(k-1)] = E\Big[\,[\mathbf{\Phi}^{n}(k-1)]\,\big\vert\,\Psi_{\mathrm{r}}^{n}(k-1)\Big]$	Estimation of $[\mathbf{\Phi}^{n}(k-1)]$ based on the current measured temperature $\Psi_{\mathrm{r}}^{n}(k-1)$
$[\mathbf{\Phi}_{\mathrm{p}}^{n}(k-1)] = E\Big[\,[\mathbf{\Phi}^{n}(k-1)]\,\big\vert\,\Psi_{\mathrm{r}}^{n}(k-2)\Big]$	Prediction of $[\mathbf{\Phi}^{n}(k-1)]$ based on the former measured temperature $\Psi_{\mathrm{r}}^{n}(k-2)$
$[\mathbf{e}_{\mathrm{e}}^{n}(k-1)] = [\mathbf{\Phi}^{n}(k-1)] - [\mathbf{\Phi}_{\mathrm{e}}^{n}(k-1)]$	Estimation error between $[\mathbf{\Phi}^{n}(k-1)]$ and $[\mathbf{\Phi}_{\mathrm{e}}^{n}(k-1)]$
$[\mathbf{e}_{\mathrm{p}}^{n}(k-1)] = [\mathbf{\Phi}^{n}(k-1)] - [\mathbf{\Phi}_{\mathrm{p}}^{n}(k-1)]$	Prediction error between $[\mathbf{\Phi}^{n}(k-1)]$ and $[\mathbf{\Phi}_{\mathrm{p}}^{n}(k-1)]$
$[\mathbf{P}_{\mathrm{e}}^{n}(k-1)] = E\Big[[\mathbf{e}_{\mathrm{e}}^{n}(k-1)]\,[\mathbf{e}_{\mathrm{e}}^{n}(k-1)]^{\mathrm{T}}\Big]$	Covariance matrix of the estimation error $[\mathbf{e}_{\mathrm{e}}^{n}(k-1)]$
$[\mathbf{P}_{\mathrm{p}}^{n}(k-1)] = E\Big[[\mathbf{e}_{\mathrm{p}}^{n}(k-1)]\,[\mathbf{e}_{\mathrm{p}}^{n}(k-1)]^{\mathrm{T}}\Big]$	Covariance matrix of the prediction error $[\mathbf{e}_{\mathrm{p}}^{n}(k-1)]$

Table 3.2. Parameters and vectors used in EKF

is from the input or observed space to an intermediate regressor vector [Bri12], and the second layer mapping is from the regressor vector to the output space. From a mathematical point of view, the two-layer mapping can be also considered as a basis function expansion, which extract the information from the original system and represent it using different combinations of basis functions.

The black-box modeling method used here is the neural network (NN) approach [JMM96]. In the temperature control system of HEPHAISTOS, two different neural networks are applied, functioning as the system estimator and the controller respectively. The NN system estimator is introduced in this section, and the NN controller will be addressed in chapter 4.

3.3.1. Introduction of Neural Network

The neural network (NN) structure was originally inspired by the concept of biological neural networks, and then gradually extended and extensively utilized in different areas. In control engineering, it is considered as a powerful alternative to conventional modeling and control methods [LV09]. It has been proven in [HSW89] that, with the appropriate network architecture, a multilayer feedforward neural network is a universal approximator for all kinds of linear and nonlinear dynamics. Due to this universal approximation ability, NN has become one of the most popular system identification methods that are applied in various applications.

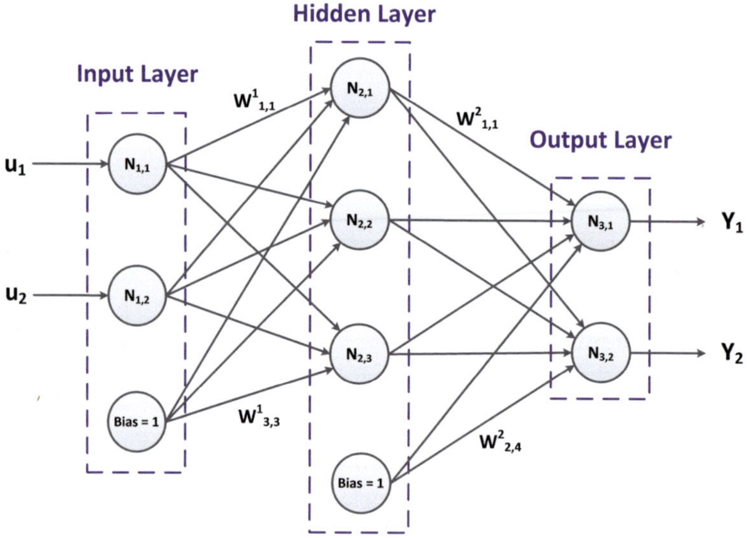

Figure 3.4. Neural network structure

A basic NN structure is illustrated in figure 3.4. Normally it consists of several elements such as nodes, synaptic links and weights. A node is also called a neuron, which is the fundamental unit in neural networks. Each node can have multiple inputs from the input of the network or other nodes, but it has only one output. The output could be either the sum of all inputs (e.g. nodes in the output layer), or the result of some local functions (perceptron) [Bar98].

The local function of each node is called the activation function, which could be either linear or nonlinear. There are different choices of activation functions (shown in figure 4.9), including the hyperbolic tangent function, softmax function, sinusoidal function, and sigmoid function [HDB+96]. It has been stated in [Hor91] that it is the architecture of the multilayer neural network not the selection of activation functions that gives the neural networks the universal approximation ability, therefore the performance difference between different activation functions is quite small. In the NN estimator and controller used in this dissertation, the hyperbolic tangent function ($\tanh$) [HDB+96] is selected (figure 3.5d), which is defined as

$$\tanh(x) = \frac{\sinh(x)}{\cosh(x)} = \frac{\mathrm{e}^{x} - \mathrm{e}^{-x}}{\mathrm{e}^{x} + \mathrm{e}^{-x}}. \tag{3.69}$$

It is selected because its derivative is easy to calculate.

The interconnection between any two nodes is called a synaptic link, or synapse, which can transfer the data from one node to another. Each synaptic link has its own weight, which can be considered as the strength of this link. One or more nodes with the same type of functions constitute a block, which is called one layer in neural networks. Besides normal nodes, in each layer there is one extra node called the bias node, whose output is always one. It never receives any data from other nodes and only transmits the output to nodes in the next layer. The function of bias nodes is to shift activation functions to different directions so that the resulted network is adaptive to different input-output dynamics [HDB+96].

In principle, a neural network comprises at least two layers, one input layer and one output layer. In most cases, a neural network also contains one or more hidden layers, which lie between the input and the output layers. Hidden layers are important for the approximation ability of neural networks. An intuitive interpretation to the function of hidden layers is that the hidden layers receive the raw information

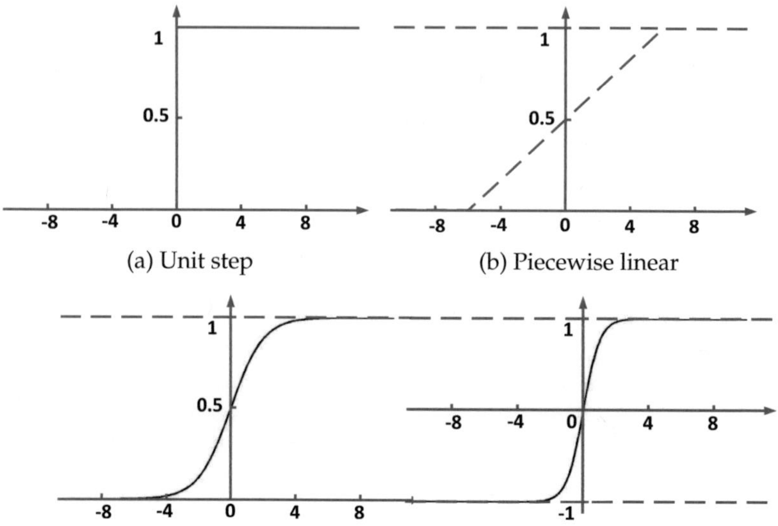

(c) Sigmoid according to [HDB$^+$96] (d) Tanh (equation 3.69)

Figure 3.5. Illustration of different activation functions.

or feature that is fed from the input layer, extract the more valuable information from the raw information, and then transmit it to the output layer for the final determination [AN15].

In other words, the task being solved by the neural network can be decomposed into a number of small subproblems, and subproblems are solved and united layer by layer. Theoretically, the neural networks with more hidden layers (also known as deep neural networks) are able to perform more complicated and accurate approximations to highly dynamic systems, and have a better performance than the so-called shallow networks that have only one or few hidden layers. However, in deep neural networks there are a number of obstacles that are still not well understood, such as the vanishing gradient problem [BSF94] or influences caused by weight initializations [SMDH13]. In practice as well as many literatures [Hay98], it is suggested to use a neural network with an appropriate number of layers instead of a deep network.

According to how different nodes are interconnected, neural networks can be divided into two types. The first type is called feedforward neural networks (FNN), as the one shown in figure 3.4, where the data is sent from the input layer, passing through hidden layers to the output layer. FNN assume that the output of the system being approximated is only determined by the input, and there is no interconnection between nodes in the same layer or feedbacks from the following layers to the former layer (m+1-th layer to m-th layer).

In contrast to FNN, there is another type of neural networks, called recurrent neural networks (RNN), where interconnections are allowed between any arbitrary nodes. The concept of RNN contains different possibilities, like the networks with feedbacks from hidden layers to the input (context) layer (like the Elman networks [Elm90]), or asynchronous fully connected networks (like the Hopfield networks [Hop82]). Feedbacks within the RNN creating internal loops or states give the network the ability to approximate more complicated systems with dynamic temporal behaviors. In principle, RNNs can be much more complex and powerful than FNNs. A RNN structure consisting of a simple feedback loop is shown in figure 3.6.

The overall function of a neural network $f(x)$ is the expansion of all activation functions. For the NN in figure 3.4, the overall function of the first output can be represented as

$$
\begin{aligned}
Y_{\mathrm{NN},1} &= f_1(u_{\mathrm{NN},1}, u_{\mathrm{NN},2}) \\
&= g_{3,1}\left(\sum_{i=1}^{3} w_{1,i}^{2} g_{2,i}\left(\sum_{j=1}^{2} w_{i,j}^{1} g_{1,j}(u_{\mathrm{NN},j}) + w_{i,3}^{1}\right) + w_{1,4}^{2}\right),
\end{aligned}
\tag{3.70}
$$

where $g_{i,j}$ is the activation function of j-th node in i-th layer, and $w_{i,j}^{l}$ represents the weight from j-th node in the l-th layer to the i-th node in the $(l+1)$-th layer (the bias node is labeled as the last node in each layer). If there is no interconnection between any two nodes, then the corresponding weight equals to zero. By adjusting the weight of each synaptic link or number of nodes in hidden layers, the behavior of the neural network is also changed. Adjustments of weights are repeated iteratively or recursively until the desired system dynamics are achieved by the network, and the corresponding process is called learning or training.

As stated in [WF05], training is the mindless kind of learning. In this dissertation, they are assumed to be equivalent. A more formal definition of learning is given in [Hay98]:

> *Learning is a process by which the free paramters of a neural network are adapted through a process of stimulation by the environment in which the network is embedded*

According to the definition, all free parameters in neural networks can be modified during learning. There are mainly two types of learning. One is similar with the process mentioned in the last paragraph, where the topology of the network is fixed and only weights are adjusted. The second type of learning involves the modification of the topology of the neural network [SM04] [EUDC07]. It increases the number of nodes in hidden layers and evolves weights of each link at the same time, according to predefined principles until the performance

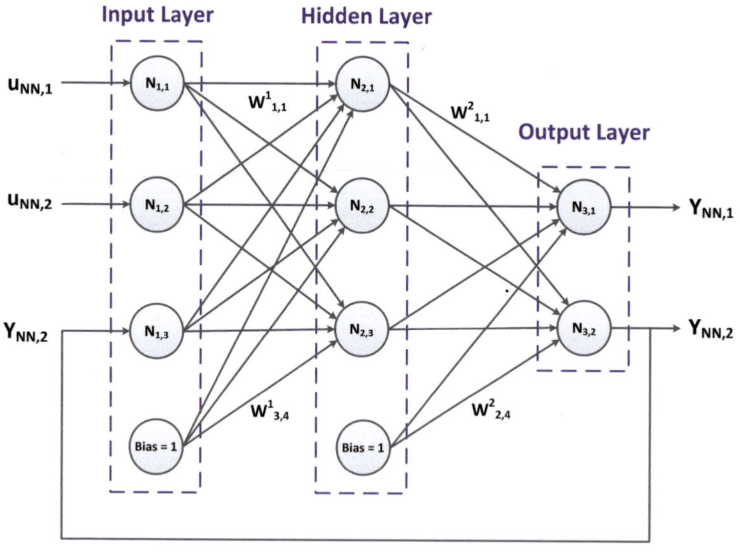

Figure 3.6. Diagram of recurrent neural network

of the network meets the learning requirement. Compared with the first fixed-topology scheme, the second learning strategy needs more computation resources and time because of the involvement of the topology adjustment [SM96]. For applications requiring fast or online learning (explained in the later section 3.3.2), the second strategy is not suitable. Therefore in this dissertation, the fixed-topology learning scheme is used. For this scheme, there are mainly three different learning approaches due to different tasks and applications.

Supervised learning

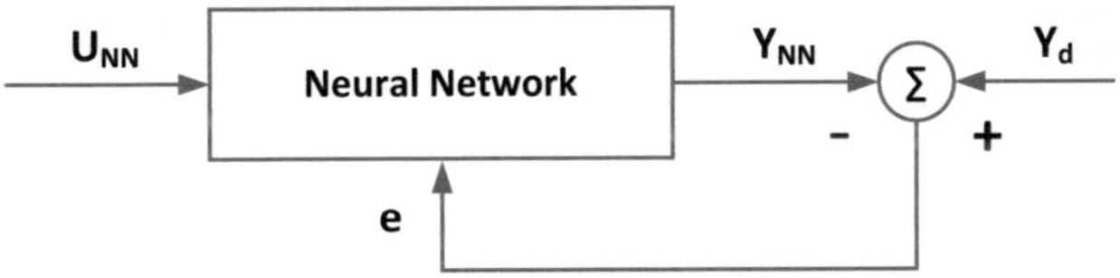

Figure 3.7. Principle of supervised learning

The name of supervised learning indicates the idea of involving a supervisor that teaches the network to learn. The principle of supervised learning is shown in figure 3.7 [HDB+96]. In supervised learning, the vector $\mathbf{U}_{\mathrm{NN}}$ is the input of the NN and $\mathbf{Y}_{\mathrm{d}}$ is the corresponding desired (correct) output vector. They are given as data pairs $(\mathbf{U}_{\mathrm{NN}}, \mathbf{Y}_{\mathrm{d}})$. The objective of supervised learning is to find the optimal (correct) weights for all links that make the output of the NN $\mathbf{Y}_{\mathrm{NN}}$ equivalent to the desired output $\mathbf{Y}_{\mathrm{d}}$ for all trained data pairs. Supervised learning is the most common learning method used in the system modeling and the system identification. Detailed introductions of supervised learning will given later in the following section 3.3.2.

Unsupervised learning

Unlike supervised learning, there is no explicit target output or supervisor regarding each input in unsupervised learning [HDB+96]. Instead of the desired output $\mathbf{Y}_{\mathrm{d}}$, for each input vector $\mathbf{U}_{\mathrm{NN}}$, the network itself decides what is the corresponding output and adjust its weights accordingly. The learning process is based on the goal of the network, normally a objective or cost function. Unsupervised learning is more used in applications of exploring statistical structure of the overall input space, such as pattern classification, decision making or data compression. In control related fields, unsupervised learning can be also applied in the controller design, where the controlling rule is discovered automatically. Unsupervised learning will be explained in the chapter 4 with more details.

Reinforcement learning

Compared with supervised and unsupervised learning, reinforcement learning is a modified combination of both [Bar98]. In reinforcement learning, for each input $\mathbf{U}_{\mathrm{NN}}$ there is also no desired output $\mathbf{Y}_{\mathrm{d}}$. Instead, here the network receives a feedback signal R from the plant (system being controlled) [Bar98], called reward or cost. The reward (cost) is not the direct output of the real controlled plant, but a real number indicating how good (bad) the corresponding input $\mathbf{U}_{\mathrm{NN}}$ is. The final objective of reinforcement learning is to choose an input policy π to maximize the overall rewards.

Reinforcement learning is essentially closer to the principle of human behaviors, which is based on the concept of learning from experiences and interactions with the environment. Compared with supervised learning, reinforcement learning is much more flexible. In theory, an optimal control rule can be obtained by using a trial-and-learn strategy [Bar98], and no concrete system model is needed in advance. More important, reinforcement learning is suitable for controlling more complicated tasks which are difficult to be described by middle-complexity mathematical models. Due to these advantages, reinforcement learning method is also used in the temperature control system of HEPHAISTOS, and detailed information will be given in chapter 4.

3.3.2. Neural Network Modeling of HEPHAISTOS

In this section, the neural network modeling of HEPHAISTOS is explained. In the beginning, the fundamental principles of supervised learning are introduced. Then the neural network structures used in HEPHAISTOS are presented. In the end, the learning (training) algorithms used in this dissertation are discussed.

Fundamentals in Supervised Learning

The training data set is represented by

$$\begin{aligned} D &= \{(\mathbf{U}_{\mathrm{NN}}(1), \mathbf{Y}_{\mathrm{d}}(1)), (\mathbf{U}_{\mathrm{NN}}(2), \mathbf{Y}_{\mathrm{d}}(2)), \ldots, (\mathbf{U}_{\mathrm{NN}}(Q), \mathbf{Y}_{\mathrm{d}}(Q))\}, \\ \mathbf{U}_{\mathrm{NN}}(q) &= \left[\, u_{\mathrm{NN},1}(q),\ u_{\mathrm{NN},2}(q),\ \ldots,\ u_{\mathrm{NN},M}(q) \,\right]^{\mathrm{T}}, \quad 1 \le q \le Q, \\ \mathbf{Y}_{\mathrm{d}}(q) &= \left[\, Y_{\mathrm{d},1}(q),\ Y_{\mathrm{d},2}(q),\ \ldots,\ Y_{\mathrm{d},N}(q) \,\right]^{\mathrm{T}}, \quad 1 \le q \le Q. \end{aligned}$$

The number Q indicates the size of the training data set D. The vector $\mathbf{U}_{\mathrm{NN}}(q)$ is the input vector and $\mathbf{Y}_{\mathrm{d}}(q)$ is the corresponding desired (correct) output vector. The target of supervised learning is to find the weight $\mathbf{w}^*$ that makes the estimated output of the network $\mathbf{Y}_{\mathrm{NN}}(q) = f(\mathbf{w}^*, \mathbf{U}_{\mathrm{NN}}(q))$ equivalent to $\mathbf{Y}_{\mathrm{d}}(q)$ for all $1 \le q \le Q$.

In order to find the best weight vector $\mathbf{w}^*$, an objective or cost function J has to be defined, which measures the error between the real and desired outputs. There are several different ways to define the cost function. For instance, a simple choice is to use the L^1-norm [HDB+96], such as

$$J_{\mathrm{L1}} = \sum_{q=1}^{Q} \left\| \mathbf{Y}_{\mathrm{NN}}(q) - \mathbf{Y}_{\mathrm{d}}(q) \right\|. \tag{3.71}$$

This L^1-norm formed cost function is intuitive, but not very commonly used in practice because the derivative is not easy to calculate. The most common cost function used in supervised learning is the following L^2-norm [HDB+96]

$$J_{\mathrm{L2}} = \min \frac{1}{2Q} \sum_{q=1}^{Q} \left(\mathbf{Y}_{\mathrm{NN}}(q) - \mathbf{Y}_{\mathrm{d}}(q)\right)^{\mathrm{T}} \left(\mathbf{Y}_{\mathrm{NN}}(q) - \mathbf{Y}_{\mathrm{d}}(q)\right), \tag{3.72}$$

which is the averaged quadratic error over all training data pair.

The principle of gradient descent is used to solve the above learning task, where the update rule for the weight vector (or matrix) is represented as [HDB+96]

$$\begin{aligned}\mathbf{w}_{\text{new}} &= \mathbf{w}_{\text{old}} + \Delta\mathbf{w} \\ &= \mathbf{w}_{\text{old}} - \eta\frac{\partial J_{\text{L2}}}{\partial \mathbf{w}},\end{aligned} \tag{3.73}$$

where η is a positive step size and $\partial J_{\text{L2}}/\partial\mathbf{w}$ is the gradient vector of $\mathbf{w}$. The weight vector (or matrix) can be defined into different forms depending on the application. For instance, in equation 3.82 weights of different layers are represented by different weight vectors. But in equation 4.21, all weights of the network are included in one weight vector. To calculate the gradient vector, all activation function with the neural network have to be differentiable. Therefore functions like the unit step function are not appropriate. Depending on different presentations of the training cost function, the learning process can be applied in two ways.

1. Batch learning: The cost function used in batch learning is exactly the same as equation 3.72, which learns from all data pairs within the set *D*. In other words, the calculation of Δw requires information from all data pairs in *D*.

2. Incremental learning: in this mode of backpropagation, the cost function is generated as [HDB+96]

$$J_{\text{IL}}(q) = \frac{1}{2}\left(\mathbf{Y}_{\text{NN}}(q) - \mathbf{Y}_{\text{d}}(q)\right)^{\text{T}}\left(\mathbf{Y}_{\text{NN}}(q) - \mathbf{Y}_{\text{d}}(q)\right),\ \ 1 \le q \le Q. \tag{3.74}$$

   The cost function 3.74 consists of only one data pair in each update. Instead of learning from all training date pairs at once, the learning data pairs in incremental learning are presented one by one to the network.

In neural network learning, one complete update of weights using the entire data set *D* is called an epoch [Hay98]. In batch learning, since the entire data set is included in the cost function (equation 3.72), each update is equivalent to one epoch, which means all weights are updated only once per epoch. In incremental learning, only one data pair is taken into account in each update. Therefore all weights are updated Q times in one epoch, starting from the first date pair $\left(\mathbf{U}_{\mathrm{NN}}(1), \mathbf{Y}(1)\right)$ until the last data pair $\left(\mathbf{U}_{\mathrm{NN}}(Q), \mathbf{Y}(Q)\right)$ is accounted for.

Online learning is a special incremental learning, with only one data pair is available at one time, therefore the cost function 3.74 can be rewritten in a online form as

$$J_{\mathrm{OL}}(k) = \frac{1}{2}\left(\mathbf{Y}_{\mathrm{NN}}(k) - \mathbf{Y}_{\mathrm{d}}(k)\right)^{\mathrm{T}}\left(\mathbf{Y}_{\mathrm{NN}}(k) - \mathbf{Y}_{\mathrm{d}}(k)\right), \tag{3.75}$$

where $\mathbf{Y}_{\mathrm{NN}}(k)$ is the estimated output from the NN at the current time k and $\mathbf{Y}_{\mathrm{d}}(k)$ is the desired (real measured) output of the system at time k. Online supervised learning is similar with the online system identification introduced in the grey-box modeling part, which is more suitable to describe the dynamics of time-varying systems.

Neural Network Modeling Approaches used in HEPHAISTOS

The implementation of neural networks in the temperature control system of HEPHAISTOS is straightforward compared with the grey-box modeling approaches. The aforementioned two learning strategies are both used to solve different tasks, as in figure 3.8.

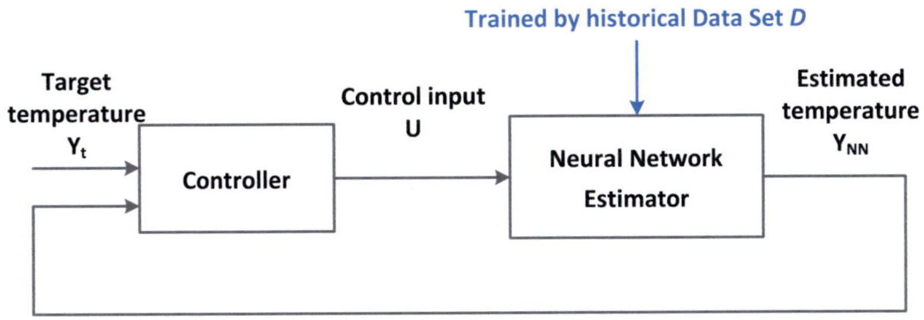

(a) Batch learning approach (offline)

In the first approach (see figure 3.8a), batch learning is used to train the NN estimator. This well-trained NN is employed as an approximation of the real plant to test the performance of different system identification algorithms or control methods. For example, in the controller design part, a controller could be firstly designed to control this NN in

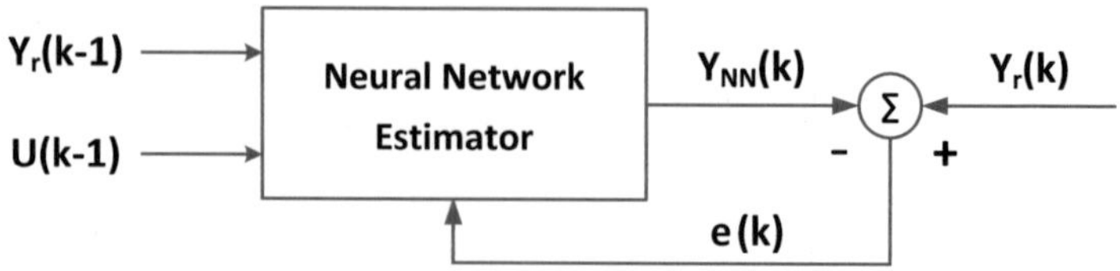

(b) Online learning approach

Figure 3.8. Neural network approaches used in this dissertation.

an offline mode. During the process of the controlling the NN estimator, the controller becomes more and more familiar with the real plant and finally it can be used to control the real plant. This method has been widely used for controller design such as in [Sch90], [WMS92], [LN95] and [GHLZ10]. In this case, an amount of historical experimental data can be used and the batch learning mode is preferred over incremental learning, due to the more stable learning results and faster converging speed.

But the limitation of this approach is that the controller trained by this NN estimator is not guaranteed to have the same performance in practice as in the test. That is because in HEPHAISTOS, the real heating process is influenced by many different factors and it is not possible to obtain experimental data that can cover all dynamics of the plant. Despite of this limitation, this NN estimator still provides valuable information and the performance on it can be considered as the benchmark to compare and select different control methods.

The second approach is to use a NN estimator for online system identification (see figure 3.8b), which functions similarly with the online system identification algorithms introduced in the grey-box modeling part. In this approach, the input of the NN estimator $\mathbf{U}_{\mathrm{NN}}$ $\big($ with the dimension $(N+M)\times 1\big)$ contains the former temperature vector $\mathbf{Y}_{\mathrm{r}}(k-1)$ and the control input vector $\mathbf{U}(k-1)$, such as

$$\mathbf{U}_{\mathrm{NN}}(k) = \begin{bmatrix} \mathbf{Y}_{\mathrm{r}}(k-1) \\ \mathbf{U}(k-1) \end{bmatrix}. \tag{3.76}$$

The output of the NN estimator $\mathbf{Y}_{\mathrm{NN}}(k)$ has the same dimension as the measured temperature vector $\mathbf{Y}_{\mathrm{r}}(k)$. The cost function is defined as in equation 3.75 with $\mathbf{Y}_{\mathrm{d}}(k) = \mathbf{Y}_{\mathrm{r}}(k)$.

Compared with the batch learning approach, the online learning approach is more suitable for the real time system identification of the plant. It has a stochastic nature that makes the weight update less likely to be trapped in a local minimum [HDB+96]. Nevertheless, online learning itself suffers from a problem that is called "catastrophic interference" or "stability-plasticity dilemma" [MC89] [Rob95]. This problem is caused by the fact that the minimum of the cost function 3.75 based on the single data pair may be far away from the minimum that is based on the whole training data set (the minimum of equation 3.72), which leads to the network completely drops all information provides by former training data pairs (prior to the current time k). In order to reduce the effect of "catastrophic interference", additional strategies are taken during the training of the NN estimator, which will be introduced in the next section.

Online Supervised Learning Algorithms in Neural Networks

The most common supervised learning algorithm used in neural networks is the backpropagation (BP) algorithm [HDB+96]. The term "backpropagation", short for the back propagation of error, refers to the process that errors at the output layer are feed back to former hidden layers and used to adjust weights accordingly. The online stochastic gradient descent backpropagation (SGBP) algorithm [HDB+96], which combines principles of online backpropagation and the aforementioned gradient descent method (see equation 3.73). is presented in this section (the detailed derivation process can be found in appendix A.2). In order to reduce the effect of "catastrophic interference", three online learning algorithms evolving from the classical online SGBP algorithm are applied in this dissertation, including the BP with momentum (BPM) algorithm, extended Kalman filter (EKF) algorithm based BP and the so-called sliding window based BP (SWBP) algorithm. Their principles are introduced in the end of this section.

Given the data set D, the complete learning process of the standard online SGBP algorithm is shown below. The notations used in the update of weights are shown in figure 3.9 and defined in table 3.3.

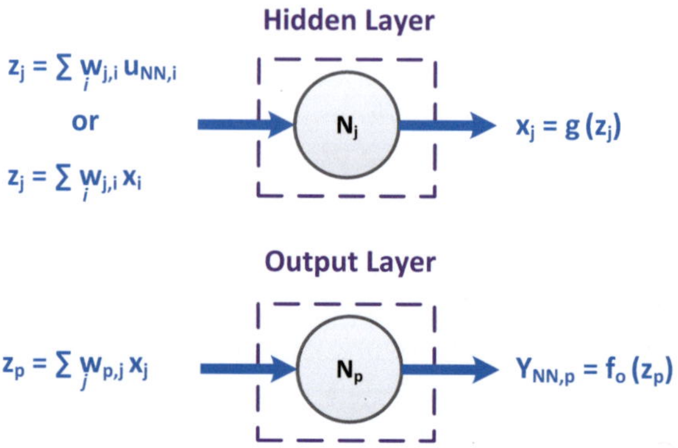

Figure 3.9. Notations used to denote the input and the output of different nodes.

1. **Initialization**:

   initialize all weights and determine activation functions f_o and g.

2. At each time k with $k \geq 2$:

   arrange the input and output of the NN estimator as

$$\mathbf{U}_{\mathrm{NN}}(k) = \begin{bmatrix} \mathbf{Y}_{\mathrm{r}}(k-1) \\ \mathbf{U}(k-1) \end{bmatrix}, \quad \mathbf{Y}_{\mathrm{d}}(k) = \mathbf{Y}_{\mathrm{r}}(k). \tag{3.77}$$

   Forward calculation: calculate the input x and output z of all nodes in this neural network using the current weights and input vector $\mathbf{U}_{\mathrm{NN}}(k)$, according to definitions in figure 3.9 and table 3.3.

   Backward error propagation: starting from the output layer, calculate the local gradient vector

$$\boldsymbol{\delta}^{\mathrm{o}}(k) = \left[\, \delta_1^{\mathrm{o}}(k),\ \delta_2^{\mathrm{o}}(k)\ \ldots\ \delta_N^{\mathrm{o}}(k) \,\right], \tag{3.78}$$

   where the element is given as [Hay98]

$$\delta_i^{\mathrm{o}}(k) = \left(Y_{\mathrm{NN},i}(k) - Y_{\mathrm{d},i}(k)\right) f_{\mathrm{o}}'(z_i(k)), \quad 1 \leq i \leq N. \tag{3.79}$$

Definition	Interpretation
N_l	Number of nodes in the l-th hidden layer ($1 \leq l \leq L$)
$x_i^l(k)$	Output of the i-th node in the l-th layer at time k
$z_j^{l+1}(k) = \sum_{i=1}^{N_l} w_{j,i}^l(k) \cdot x_i^l(k)$	Input of the j-th node in the $(l+1)$-th layer at time k
$\delta_j^l(k) = \frac{\partial J_{\mathrm{OL}}(k)}{\partial z_j^l(k)}$	Local gradient of the j-th node in the l-th layer at time k
$f_\mathrm{o},\ f_\mathrm{o}'(z) := \frac{\mathrm{d}f_\mathrm{o}'}{\mathrm{d}z}$	Activation function used in the output layer and its first derivative
$g,\ g'(z) := \frac{\mathrm{d}g'}{\mathrm{d}z}$	Activation function used in the hidden layers and its first derivative

Table 3.3. Parameters used in the update of weights (in NN).

The values $Y_{\mathrm{NN},i}(k)$ and $Y_{\mathrm{d},i}(k)$ are the i-th element from the vector $\mathbf{Y}_{\mathrm{NN}}(k)$ and $\mathbf{Y}_\mathrm{d}(k)$, respectively.

For the l-th hidden layer ($1 \leq l \leq L$), the corresponding local gradient vector is [Hay98]

$$\left[\boldsymbol{\delta}^l(k)\right] = \left[\, \delta_1^l(k),\ \delta_2^l(k),\ \ldots\ \delta_{N_l}^l(k) \,\right], \tag{3.80}$$

where the element is defined as

$$\delta_i^l(k) = g'(z_i^l(k)) \cdot \sum_{j=1}^{N_{l+1}} \delta_j^{l+1}(k)\, w_{j,i}^{l+1}(k-1), \quad 1 \leq i \leq N_l. \tag{3.81}$$

with

$$\left[\mathbf{w}^{l+1}(k-1)\right] = \begin{bmatrix} w_{1,1}^{l+1}(k-1) & w_{1,2}^{l+1}(k-1) & \cdots & w_{1,N_l}^{l+1}(k-1) \\ w_{2,1}^{l+1}(k-1) & w_{2,2}^{l+1}(k-1) & \cdots & w_{2,N_l}^{l+1}(k-1) \\ \vdots & \vdots & \ddots & \vdots \\ w_{N_{l+1},1}^{l+1}(k-1) & w_{N_{l+1},2}^{l+1}(k-1) & \cdots & w_{N_{l+1},N_l}^{l+1}(k-1) \end{bmatrix}. \tag{3.82}$$

The $L+1$-th hidden layer is defined as the output layer ($\left[\mathbf{w}^{L+1}(k-1)\right] = \left[\mathbf{w}^{\mathrm{o}}(k-1)\right], N_{l+1} = N_{\mathrm{o}}$).

Weight update: for weights of links connecting to the output layer, the update rule is given by [Hay98]

$$\left[\mathbf{w}^{\mathrm{o}}(k)\right] = \left[\mathbf{w}^{\mathrm{o}}(k-1)\right] - \eta \cdot \left(\mathbf{x}^{L}(k)\boldsymbol{\delta}^{\mathrm{o}}(k)\right)^{\mathrm{T}}, \tag{3.83}$$

with the output vector is defined as

$$\mathbf{x}^{L}(k) = \left[\, x_1^L(k),\; x_2^L(k),\; \ldots,\; x_{N_L}^L(k)\,\right]^{\mathrm{T}}. \tag{3.84}$$

For weights of links only connecting with hidden layers, the update rule is given by [Hay98]

$$\left[\mathbf{w}^{l}(k)\right] = \left[\mathbf{w}^{l}(k-1)\right] - \eta \cdot \left(\mathbf{x}^{l-1}(k)\boldsymbol{\delta}^{l}(k)\right)^{\mathrm{T}}, \quad 1 \le l \le L \tag{3.85}$$

with the output vector is defined as

$$\mathbf{x}^{l}(k) = \left[\, x_1^l(k),\; x_2^l(k),\; \ldots,\; x_{N_l}^l(k)\,\right]^{\mathrm{T}}. \tag{3.86}$$

When the value $l = 1$, the vector $\mathbf{x}^{l-1}(k)$ is defined as the input vector of the NN

$$\mathbf{x}^{0}(k) = \mathbf{U}_{\mathrm{NN}}(k). \tag{3.87}$$

3. Procedure 2 is executed once per time step.

There are mainly two rounds of calculations in the backpropagation algorithm. One is the forward pass which calculate the output of each node from the input layer to the output layer. The other one is the backward error propagation which computes the local gradient of each node from the output layer to the input layer. After these two rounds of calculations, all weights can be updated. Intuitively, the forward pass provides necessary information for the backward pass, and the backward pass calculates how much of the final output error belongs to each individual weights, which will be updated accordingly. The backpropagation algorithm reduces the overall computation cost significantly and promotes the development of neural networks a big step forward.

Based on the standard online SGBP algorithm, additional strategies have to be taken to reduce the effect of "catastrophic interference". As stated in [Fre99], there are several possible strategies, by either using noise-free training data, applying more constructive network architectures or adding extra information storage besides pure weights of the network. The first two approaches are difficult to implement in realistic applications, therefore most practical online learning algorithms go to the third direction, such as the following three algorithms.

- BPM is also named as enhanced backpropagation [Roj96], which involves the momentum term in the weight update expression, such as

$$\Delta \mathbf{w}(q) = -\eta \left. \frac{\partial J_{\mathrm{OL}}}{\partial \mathbf{w}} \right|_k + \alpha \, \Delta \mathbf{w}(q-1), \tag{3.88}$$

  where α is a tuning factor indicating the influence from the momentum. The involvement of momentum brings information and influences from past training date pairs, which effectively attenuate weight oscillations during the iteration process and improves the estimation stability.

- This idea of using EKF in neural networks learning was firstly applied in [SW89] and quickly became a powerful and efficient learning algorithm, especially in incremental learning mode [KV92] [RRK$^+$92] [WVDM00]. The update equations using EKF in NN training is similar with the procedures described in former nonlinear system identification (equations 3.67 and 3.68), and the corresponding derivation process can also be found in [TP08].

- Sliding window based backpropagation [CMA94] is a way to use batch learning mode in a online form. Instead of using the entire training set D as in classical batch learning, the basic idea is that at each time k, the former N_s data pairs (sliding window) are used for the training (from time $k - N_s + 1$ to the current time k). Therefore the cost function is changed into

$$J_{\mathrm{SW}}(k) = \frac{1}{2N_s} \sum_{q=k-N_s+1}^{k} \Big(\mathbf{Y}(q) - \mathbf{Y}_{\mathrm{NN}}(q)\Big)^{\mathrm{T}} \Big(\mathbf{Y}(q) - \mathbf{Y}_{\mathrm{NN}}(q)\Big). \tag{3.89}$$

At each time k, not only the current observed data pair but also N_s former data pairs are included in this cost function $J_{\mathrm{SW}}(k)$. The similar idea is widely used in neural network based time series predictions, such as in [FDH01].

These three algorithms are all fast and efficient online NN training algorithms that are implemented in practical applications. They have been tested and compared in real experiments of HEPHAISTOS, and corresponding results can be found in chapter 5. Further details about the neural network design and implementations refer to [HDB+96], [Roj96], [Hay98] and [AN15].

4. Control System Design

Controller design is the final and most important part in the temperature control system. For microwave heating systems with multiple distributed feeding sources as implemented in HEPHAISTOS, several factors have to be considered before the selection of appropriate control algorithms:

- Highly coupled dynamics;
- High input and output dimensions;
- Linear and nonlinear systems with modeling uncertainties;
- Time-varying model parameters;
- Input constraints.

A number of currently main stream control methods are briefly analyzed below, with respect to aforementioned properties:

- MIMO PID [MKJ02] : PID is a powerful control scheme that is mostly applied in SISO systems. The so-called MIMO PID is its extension to MIMO systems. The applicable systems are mostly systems that can be decentralized such as in [XCH06] and [PKM13], or small scale fully coupled MIMO systems, for example, 2×2 MIMO systems [RLRJGA06] [CC14]. For non-decoupled MIMO systems with high input and output dimensions, the tuning of MIMO PID parameters will become complex and inefficient.
- Linear quadratic regulator (LQR) [AM07]: LQR is one of the most notable controllers that have been widely used in the last few decades. The objective of LQR is to optimize a cost function based on future system behaviors. Thus essentially it is a predictive control method. The control action of LQR comes from the solution of a discrete-time algebraic Riccati equation (DARE) [LR95], which is

a nonlinear equation normally solved by numerical methods. Although LQR has a similar control principle as MPC, various results [GBTH06] [FKÖS11] demonstrated that with equal or even more control effort, the performance of LQR is not as good as MPC. It was also suggested in [EDH05] that for MIMO systems which are not fully decoupled, e.g. HEPHAISTOS, MPC is more suitable.

- H_2 and H_∞ control [Bur98]: both H_2 and H_∞ control are powerful and robust methods that can be used for systems with noise and model uncertainties [SP07]. However, high computation costs is still the obstacle that limits their large scale applications. In many cases it is difficult to use these two methods to get analytical solutions, and they can only be solved numerically [Bur98]. Moreover, both H_2 and H_∞ control methods are still too complicated to be applied in high dimensional MIMO systems [vdB07].
- Fuzzy logic control (FLC) [Lee90]: FLC is a control method which is widely applied in systems with unknown dynamics, benefiting from empirical knowledge and experience. For SISO or MIMO systems with low dimensions, FLC is easily implemented and tuned. But for MIMO systems with high dimensions, the numbers of both fuzzy rules and membership functions are increasing exponentially, which rises the difficulties in the controller design and the tuning of FLC [MSH98]. In addition, for MIMO systems with highly coupled dynamics, empirical knowledge is difficult to be directly used for the generation of fuzzy rules, which leads to a significant performance degradation of FLC.

Considering above reasons and analyses, two control schemes are introduced in this chapter. The first adaptive control scheme includes the model predictive control (MPC) [MHL99] and neural network control (NNC) methods [CK92], aiming to control temperatures of separated points. The other intelligent control scheme is based on the reinforcement learning control (RLC) method [Bar98]. Instead of individual temperatures, it directly controls characteristics of the heating pattern. Both control schemes are optimized according to properties of HEPHAISTOS, to reduce the computation complexity and improve the control performance.

4.1. Adaptive Control

In general, the control methods that adjust their control actions based on the varying parameters of the controlled systems are called adaptive control [KV86]. They can be divided into two categories, indirect and direct adaptive control. The indirect control scheme contains online system identification to describe the system dynamics, using measured system data. In direct adaptive control, no online system identification is needed and the control action relies on direct feedbacks from the plant, such as the most well-known PID controller. Indirect control schemes are more suitable for systems modeled with unknown constants and slow-varying parameters, while direct control schemes are mainly used to deal with systems with quickly varying dynamics. In HEPHAISTOS, both control schemes have been applied in the temperature control system: model predictive control (MPC) as an indirect adaptive control scheme and neural network based control (NNC) as a direct/semi-direct adaptive control. Details of MPC and NNC are introduced in this section.

4.1.1. Model Predictive Control

Model predictive control (MPC), oriented from the receding horizon control (RHC) [MHL99], is an advanced and well-developed feedback control method. It was firstly implemented in the oil industry [RRTP78] and then spread to other fields like chemistry and aerospace [QB03]. In this section, the fundamentals of MPC are introduced at first, and then details of MPC implementations regarding two former derived models (equation 3.41 and 3.46) are presented (linear MPC and nonlinear MPC).

No matter which type of models MPC is applied on, the control steps are the same, which are according to [CA13]

- Predict the future system output based on the given or estimated system parameters.
- Calculate a control sequence that minimizes the predefined objective function.

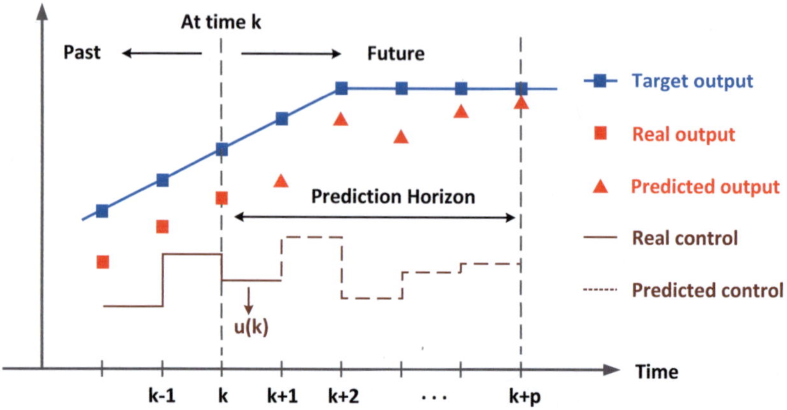

Figure 4.1. Principle of MPC control algorithms.

- Apply the first control action of the sequence as the real control input and drop the rest.
- Repeat above procedures at each time instants.

The idea of MPC is illustrated by figure 4.1. Further elaborations of MPC can be found in [CBCB04] and [MHL99].

MPC has natural advantages over other control methods (mentioned in the beginning of this chapter) in dealing with complex systems like HEPHAISTOS:

- The principle is intuitive and easy implement.
- The state-space model formed control law is suitable when the input and output dimensions are large.
- A great variety of dynamics including delay times and external disturbances can be easily handled.
- It involves predictions for the future behaviors which are useful for tracking time-varying trajectories.
- It can easily be extended to treat different kinds of input and output constraints.

Due to above reasons, MPC can be used in the temperature control system of HEPHAISTOS. The entire structure of the MPC based temperature control system is shown as in figure 4.2.

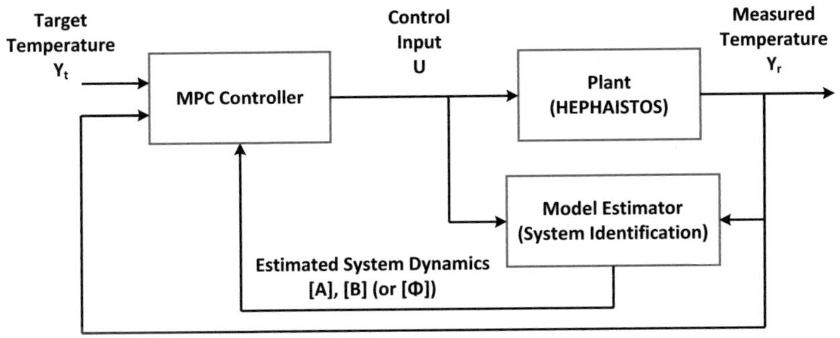

Figure 4.2. Temperature control system of HEPHAISTOS using MPC.

In this temperature control system, there are two key components that determine the final control performance.

1. Model estimator

   The model estimator is used to capture the dynamics of the HEPHAISTOS oven and send the estimated model dynamics, e.g. matrices $[\mathbf{A}(k)]$, $[\mathbf{B}(k)]$ or $[\mathbf{\Phi}^n(k)]$, to the MPC controller. The accuracy of the estimated system model directly influences the final control performance of any MPC algorithms, and the form of the estimated model also determines which kind of MPC algorithms is appropriate for the given application.

   In principle, both the grey-box and black-box modeling approaches introduced in the last chapter could be used as the system estimator in the above figure 4.2. But in practice, the evaluation of the NN model takes much more time than that of grey-box models. In order to keep a fast control pace, only the grey-box approaches (linear 3.41 and nonlinear 3.46 models) are implemented in this model estimator. The NN model will be used in the neural network controller in section 4.1.2.

2. MPC controller

   The MPC controller is responsible for calculating the new control input vector and giving it to the HEPHAISTOS oven, according to the estimated model and the MPC principle.

   In the MPC controller, the first task is to define the cost function, which is the objective to be solved in the controller. It is assumed that the modified target temperature vector at time k is

$$\mathbf{Y}_{\mathrm{t}}(k) = \mathbf{T}_{\mathrm{t}}(k) - \mathrm{T}_{\mathrm{a}}(k), \tag{4.1}$$

   where $\mathbf{T}_{\mathrm{t}}(k)$ is the real target temperature vector $(N \times 1)$ and $\mathrm{T}_{\mathrm{a}}(k)$ is the temperature of the surrounding air. One of the most common cost functions $J(k)$ used in the MPC method is [Wan09]

$$\begin{aligned} J(k) = \sum_{l=1}^{p} \Big[\big(\mathbf{Y}_{\mathrm{t}}(k+l) - \mathbf{Y}(k+l)\big)^{\mathrm{T}} \mathbf{\Gamma}(l) \big(\mathbf{Y}_{\mathrm{t}}(k+l) - \mathbf{Y}(k+l)\big) \\ + \mathbf{U}^{\mathrm{T}}(k+l-1) \mathbf{\Lambda}(l) \mathbf{U}(k+l-1) \Big], \end{aligned} \tag{4.2}$$

   with

$$\begin{aligned} \mathbf{Y}(l) &= \begin{bmatrix} Y_1(l) \\ Y_2(l) \\ \vdots \\ Y_N(l) \end{bmatrix}, \quad [\mathbf{\Gamma}(l)] = \begin{bmatrix} \gamma(l) & 0 & \cdots & 0 \\ 0 & \gamma(l) & \cdots & 0 \\ \vdots & \vdots & \ddots & \vdots \\ 0 & 0 & \cdots & \gamma(l) \end{bmatrix}, \\ \mathbf{U}(l) &= \begin{bmatrix} u_1(l) \\ u_2(l) \\ \vdots \\ u_M(l) \end{bmatrix}, \quad [\mathbf{\Lambda}(l)] = \begin{bmatrix} \Lambda(l) & 0 & \cdots & 0 \\ 0 & \Lambda(l) & \cdots & 0 \\ \vdots & \vdots & \ddots & \vdots \\ 0 & 0 & \cdots & \Lambda(l) \end{bmatrix}, \end{aligned} \tag{4.3}$$

   where $\mathbf{\Gamma}(l)$ $(N \times N)$ and $\mathbf{\Lambda}(l)$ $(M \times M)$ are both positive definite tuning matrices used to adjust weights of future temperature and control input vectors, respectively. The future temperature vectors can be predicted using either the linear model 3.41 (linear MPC) or the nonlinear model 3.46 (nonlinear MPC).

In the above cost function p denotes the prediction and control horizon length. In general, the prediction horizon length is not equivalent to the control horizon length. But here in order to simplify the derivation, it is assumed they are identical, which will not affect the control performance.

Based on the cost function, the formal representation of the control task can be expressed as

$$\begin{aligned} &\min \quad J(k) \\ &\text{subject to} \quad 0 \leq u_m(k+l-1) \leq 1, \quad \text{for } 1 \leq m \leq M,\ 1 \leq l \leq p, \end{aligned} \tag{4.4}$$

where the constraints in equation 4.4 guarantees that the microwave feeding power is always between 0 and 100%. This control task consists of two aspects. The first aspect is to track the predefined target temperature and minimize the quadratic difference between the measured temperatures and the target temperature (the first term in the cost function $J(k)$). The second aspect refers to the minimization of the control power and save the heating energy (the second term in the cost function $J(k)$).

Defining a control sequence (as illustrated in figure 4.1)

$$U_{\text{set}}(k) = \{\mathbf{U}(k),\ \mathbf{U}(k+1),\ \ldots,\ \mathbf{U}(k+p-1)\}, \tag{4.5}$$

the control solution $\mathbf{U}(k)$ can be obtained by searching for the optimal sequence $U^*_{\text{set}}(k)$ fulfilling

$$U^*_{\text{set}}(k) = \arg_U \min J(k), \tag{4.6}$$

and then implementing the first control input vector from the optimal sequence. When the controlled system model is linear (such as equation 3.41), normally an analytical expression can be derived for $U^*_{\text{set}}(k)$ as well as $\mathbf{U}(k)$. Otherwise when the system model is nonlinear (such as equation 3.46), an analytical control solution is not feasible. In this case, the control task 4.4 has to be solved with a numerical solution.

In the following, the linear MPC and the nonlinear MPC methods will be introduced in detail.

Linear MPC

In this section, procedures of how the control solution is derived in the linear MPC system is presented. Linear MPC achieves great successes in a number of applications, e.g. in [QB03] and [MHL99]. By now, the linear MPC theory is widely implemented, and over 90% of MPC applications are linear. Various linear MPC algorithms exist, including the dynamic matrix control (DMC) [Wan09], predictive functional control (PFC) [HBS11], model algorithmic control (MAC) [GPM89], and generalized predictive control (GPC) [CMT87]. Among these different algorithms, DMC is one of the most powerful and widely implemented MPC algorithms, especially in industry. According to the investigation in [OOH95], all major oil companies apply DMC-like approaches to control process variables such as the temperature and pressure in their new installations or revamps. Compared with other MPC approaches like GPC, DMC is more suitable for multivariable state-space formed systems as well as ARX types of models. In order to further involve the control input constraints, an improved version of DMC - quadratic DMC (QDMC) [GM86] algorithm is applied.

The time-invariant QDMC algorithm is not designed to deal with time-varying systems (such as equation 4.7), but still it is rational to be implemented here. This algorithm is suitable because the currently estimated system model (at time k) can predict the future behavior (until time $k + p$) accurately. In microwave heating applications, the system model parameters are changing slowly, especially when the temperature of the load is varying within a small range (when the target temperature is fixed). As long as the current system parameters are estimated accurately, it is reasonable to assume that the current estimated parameters can be used to predict future outputs with a high accuracy. For instance, it has been demonstrated in [NP97] that the time-invariant MPC approach can keep a suboptimal performance in practical time-varying applications. On the other hand, although a number of MPC approaches have been proposed specialized for specific time-varying systems, such as the approaches in [DC03] [ZL03] and [Ric05], the performance gain of these approaches over conventional time-invariant MPC approaches is limited. In this thesis, the time-invariant QDMC algorithm is selected.

QDMC contains two main steps. The first step is to apply the standard DMC to the control task and obtain a control solution. If the control solution fulfills the input constraint (as given by equation 4.4), it will be directly used as the true control solution $\mathbf{U}(k)$. Otherwise if it does not fulfill the input constraint, a new quadratic programming (QP) problem [GM86] has to be formulated and another control solution will be calculated. Procedures of QDMC are given in the following.

In standard time-invariant DMC, the linear discrete-time equation 3.41 is firstly re-written in a complete state-space form as [Wan09]

$$\begin{aligned} \mathbf{X}(k+1) = & \ [\mathbf{A}(k)]\,\mathbf{X}(k) + [\mathbf{B}(k)]\,\mathbf{U}(k), \\ \mathbf{Y}(k) = & \ \mathbf{X}(k), \end{aligned} \tag{4.7}$$

where $\mathbf{X}(k)$ is the state variable vector. Defining the difference state, input and output vectors as

$$\begin{aligned} \Delta\mathbf{X}(k) &= \mathbf{X}(k) - \mathbf{X}(k-1), \\ \Delta\mathbf{U}(k) &= \mathbf{U}(k) - \mathbf{U}(k-1), \\ \Delta\mathbf{Y}(k) &= \mathbf{Y}(k) - \mathbf{Y}(k-1), \end{aligned}$$

there is

$$\Delta\mathbf{X}(k+1) = \ \mathbf{X}(k+1) - \mathbf{X}(k) = \ [\mathbf{A}(k)]\,\Delta\mathbf{X}(k) + [\mathbf{B}(k)]\,\Delta\mathbf{U}(k). \tag{4.8}$$

In order to simplify notations during following derivations, new augmented matrices are defined as [Wan09]

$$\mathbf{X}_{\mathrm{d}}(k) = \begin{bmatrix} \Delta\mathbf{X}(k) \\ \mathbf{Y}(k) \end{bmatrix}, \qquad [\mathbf{A}_{\mathrm{d}}(k)] = \begin{bmatrix} [\mathbf{A}(k)] & [\mathbf{O}_N] \\ [\mathbf{A}(k)] & [\mathbf{I}_N] \end{bmatrix},$$

$$[\mathbf{B}_{\mathrm{d}}(k)] = \begin{bmatrix} [\mathbf{B}(k)] \\ [\mathbf{B}(k)] \end{bmatrix}, \qquad [\mathbf{C}_{\mathrm{d}}(k)] = \begin{bmatrix} [\mathbf{O}_N] & [\mathbf{I}_N] \end{bmatrix},$$

where $[\mathbf{O}_N]$ and $[\mathbf{I}_N]$ are zero and identity matrices ($N \times N$).

Substituting above new defined matrices into equation 4.8, a new state-space formed system model is generated as

$$\begin{aligned} \mathbf{X}_{\mathrm{d}}(k+1) = & \ [\mathbf{A}_{\mathrm{d}}(k)]\,\mathbf{X}_{\mathrm{d}}(k) + [\mathbf{B}_{\mathrm{d}}(k)]\,\Delta\mathbf{U}(k), \\ \mathbf{Y}(k) = & \ [\mathbf{C}_{\mathrm{d}}(k)]\,\mathbf{X}_{\mathrm{d}}(k) \end{aligned} \tag{4.9}$$

In time-invariant DMC and QDMC, the future temperatures can be predicted (until the time $k+p$) using the the current temperature and the current estimated parameters, such as in [Wan09] ($1 \leq l \leq p$)

$$\begin{aligned}\mathbf{Y}(k+l) &= [\mathbf{C}_\mathrm{d}(k)]\,\mathbf{X}_\mathrm{d}(k+l)\\ &= [\mathbf{C}_\mathrm{d}(k)]\Big([\mathbf{A}_\mathrm{d}(k)]\,\mathbf{X}_\mathrm{d}(k+l-1)+[\mathbf{B}_\mathrm{d}(k)]\,\Delta\mathbf{U}(k+l-1)\Big).\end{aligned} \tag{4.10}$$

All future temperature vectors from $k+1$ to $k+p$ can be generated using this equation

$$\begin{aligned}\mathbf{Y}(k+1\,|\,k) &= [\mathbf{C}_\mathrm{d}(k)]\,[\mathbf{A}_\mathrm{d}(k)]\,\mathbf{X}_\mathrm{d}(k)+[\mathbf{C}_\mathrm{d}(k)]\,[\mathbf{B}_\mathrm{d}(k)]\,\Delta\mathbf{U}(k),\\ \mathbf{Y}(k+2\,|\,k) &= [\mathbf{C}_\mathrm{d}(k)]\,[\mathbf{A}_\mathrm{d}(k)]^2\,\mathbf{X}_\mathrm{d}(k)+[\mathbf{C}_\mathrm{d}(k)]\,[\mathbf{A}_\mathrm{d}(k)]\,[\mathbf{B}_\mathrm{d}(k)]\,\Delta\mathbf{U}(k)\\ &\quad+[\mathbf{C}_\mathrm{d}(k)]\,[\mathbf{B}_\mathrm{d}(k)]\,\Delta\mathbf{U}(k+1),\\ &\;\;\vdots\\ \mathbf{Y}(k+p\,|\,k) &= [\mathbf{C}_\mathrm{d}(k)]\,[\mathbf{A}_\mathrm{d}(k)]^p\,\mathbf{X}_\mathrm{d}(k)\\ &\quad+[\mathbf{C}_\mathrm{d}(k)]\,[\mathbf{A}_\mathrm{d}(k)]^{p-1}\,[\mathbf{B}_\mathrm{d}(k)]\,\Delta\mathbf{U}(k)+\cdots\\ &\quad+[\mathbf{C}_\mathrm{d}(k)]\,[\mathbf{B}_\mathrm{d}(k)]\,\Delta\mathbf{U}(k+p-1).\end{aligned} \tag{4.11}$$

The whole system can be represented in a compact form as [Wan09]

$$\mathbf{Y}_\mathrm{d}(k) = [\mathbf{F}_\mathrm{d}(k)]\,\mathbf{X}_\mathrm{d}(k)+[\boldsymbol{\Xi}_\mathrm{d}(k)]\,\Delta\mathbf{U}_\mathrm{d}(k), \tag{4.12}$$

with

$$[\boldsymbol{\Xi}_\mathrm{d}(k)] = \begin{bmatrix} [\mathbf{C}_\mathrm{d}(k)]\,[\mathbf{B}_\mathrm{d}(k)] & [\mathbf{O}_{N\times M}] & \cdots & [\mathbf{O}_{N\times M}]\\ \begin{matrix}[\mathbf{C}_\mathrm{d}(k)]\,[\mathbf{A}_\mathrm{d}(k)]\\ \cdot\,[\mathbf{B}_\mathrm{d}(k)]\end{matrix} & [\mathbf{C}_\mathrm{d}(k)]\,[\mathbf{B}_\mathrm{d}(k)] & \cdots & [\mathbf{O}_{N\times M}]\\ \vdots & \vdots & \ddots & \vdots\\ \begin{matrix}[\mathbf{C}_\mathrm{d}(k)]\,[\mathbf{A}_\mathrm{d}(k)]^{p-1}\\ \cdot\,[\mathbf{B}_\mathrm{d}(k)]\end{matrix} & \begin{matrix}[\mathbf{C}_\mathrm{d}(k)]\,[\mathbf{A}_\mathrm{d}(k)]^{p-2}\\ \cdot\,[\mathbf{B}_\mathrm{d}(k)]\end{matrix} & \cdots & [\mathbf{C}_\mathrm{d}(k)]\,[\mathbf{B}_\mathrm{d}(k)] \end{bmatrix},$$

$$[\mathbf{F}_\mathrm{d}(k)] = \begin{bmatrix} [\mathbf{C}_\mathrm{d}(k)]\,[\mathbf{A}_\mathrm{d}(k)] \\ [\mathbf{C}_\mathrm{d}(k)]\,[\mathbf{A}_\mathrm{d}(k)]^2 \\ [\mathbf{C}_\mathrm{d}(k)]\,[\mathbf{A}_\mathrm{d}(k)]^3 \\ \vdots \\ [\mathbf{C}_\mathrm{d}(k)]\,[\mathbf{A}_\mathrm{d}(k)]^p \end{bmatrix}, \quad \mathbf{Y}_\mathrm{d}(k) = \begin{bmatrix} \mathbf{Y}(k+1|k) \\ \mathbf{Y}(k+2|k) \\ \mathbf{Y}(k+3|k) \\ \vdots \\ \mathbf{Y}(k+p|k) \end{bmatrix},$$

$$\Delta\mathbf{U}_\mathrm{d}(k) = \begin{bmatrix} \Delta\mathbf{U}(k) \\ \Delta\mathbf{U}(k+1) \\ \Delta\mathbf{U}(k+2) \\ \vdots \\ \Delta\mathbf{U}(k+p-1) \end{bmatrix}, \tag{4.13}$$

where $[\mathbf{O}_{N\times M}]$ is the zero matrix with the dimension $N \times M$.

With the modified state-space system model 4.12, the original cost function 4.2 is also modified into [Wan09]

$$J_\mathrm{LM}(k) = \Big[\Big(\mathbf{Y}_\mathrm{d,t}(k) - \mathbf{Y}_\mathrm{d}(k)\Big)^\mathrm{T} [\mathbf{\Gamma}] \Big(\mathbf{Y}_\mathrm{d,t}(k) - \mathbf{Y}_\mathrm{d}(k)\Big) \\ + \Delta\mathbf{U}_\mathrm{d}(k)^\mathrm{T} [\mathbf{\Lambda}]\, \Delta\mathbf{U}_\mathrm{d}(k) \Big] \tag{4.14}$$

$$= \Big[\Big(\mathbf{Y}_\mathrm{d,t}(k) - [\mathbf{F}_\mathrm{d}(k)]\,\mathbf{X}_\mathrm{d}(k)\Big)^\mathrm{T} \mathbf{\Gamma} \Big(\mathbf{Y}_\mathrm{d,t}(k) - [\mathbf{F}_\mathrm{d}(k)]\,\mathbf{X}_\mathrm{d}(k)\Big) \\ - 2\Delta\mathbf{U}_\mathrm{d}^\mathrm{T}(k)\,[\mathbf{\Xi}_\mathrm{d}(k)]^\mathrm{T}\,[\mathbf{\Gamma}]\,\Big(\mathbf{Y}_\mathrm{d,t}(k) - [\mathbf{F}_\mathrm{d}(k)]\,\mathbf{X}_\mathrm{d}(k)\Big) \\ + \Delta\mathbf{U}_\mathrm{d}^\mathrm{T}(k)\Big(\,[\mathbf{\Xi}_\mathrm{d}(k)]^\mathrm{T}\,[\mathbf{\Gamma}]\,[\mathbf{\Xi}_\mathrm{d}(k)] + [\mathbf{\Lambda}]\,\Big)\Delta\mathbf{U}_\mathrm{d}(k)\Big], \tag{4.15}$$

with

$$[\mathbf{\Gamma}] := \begin{bmatrix} [\mathbf{\Gamma}(1)] & [\mathbf{O}_N] & \cdots & [\mathbf{O}_N] \\ [\mathbf{O}_N] & [\mathbf{\Gamma}(2)] & \cdots & [\mathbf{O}_N] \\ \vdots & \vdots & \ddots & \vdots \\ [\mathbf{O}_N] & [\mathbf{O}_N] & \cdots & [\mathbf{\Gamma}(p)] \end{bmatrix},$$

$$[\boldsymbol{\Lambda}] := \begin{bmatrix} [\boldsymbol{\Lambda}(1)] & [\mathbf{O}_M] & \cdots & [\mathbf{O}_M] \\ [\mathbf{O}_M] & [\boldsymbol{\Lambda}(2)] & \cdots & [\mathbf{O}_M] \\ \vdots & \vdots & \ddots & \vdots \\ [\mathbf{O}_M] & [\mathbf{O}_M] & \cdots & [\boldsymbol{\Lambda}(p)] \end{bmatrix},$$

and $\mathbf{Y}_{\mathrm{d,t}}(k)$ is the modified target temperature vector as

$$\mathbf{Y}_{\mathrm{d,t}}(k) = \left[\mathbf{Y}_{\mathrm{t}}(k)^{\mathrm{T}} \ \mathbf{Y}_{\mathrm{t}}(k+1)^{\mathrm{T}} \ \ldots \ \mathbf{Y}_{\mathrm{t}}(k+p-1)^{\mathrm{T}}\right]^{\mathrm{T}}.$$

Both $[\boldsymbol{\Gamma}]$ $(Np \times Np)$ and $[\boldsymbol{\Lambda}]$ $(Mp \times Mp)$ are positive definite diagonal matrices that are used to adjust weights of $\Delta\mathbf{U}(k)$ and $\mathbf{Y}_{\mathrm{d,t}}(k)$ respectively. Substituting the equations 4.12 and 4.13 into the above cost function and differentiating both sides with respect to $\Delta\mathbf{U}_{\mathrm{d}}(k)$, the analytical formed solution is given as [Wan09]

$$\begin{aligned} \Delta\mathbf{U}_{\mathrm{d}}(k) =& \left([\boldsymbol{\Xi}_{\mathrm{d}}(k)]^{\mathrm{T}} [\boldsymbol{\Gamma}]^{\mathrm{T}} [\boldsymbol{\Gamma}] [\boldsymbol{\Xi}_{\mathrm{d}}(k)] + [\boldsymbol{\Lambda}]\right)^{-1} [\boldsymbol{\Xi}_{\mathrm{d}}(k)]^{\mathrm{T}} [\boldsymbol{\Gamma}]^{\mathrm{T}} [\boldsymbol{\Gamma}] \\ & \left(\mathbf{Y}_{\mathrm{d,t}}(k) - [\mathbf{F}_{\mathrm{d}}(k)]\, \mathbf{X}_{\mathrm{d}}(k)\right), \\ \mathbf{U}(k) =& \ \mathbf{U}(k-1) + \Delta\mathbf{U}(k), \end{aligned} \tag{4.16}$$

where $\Delta\mathbf{U}(k)$ is the first vector taken from $\Delta\mathbf{U}_{\mathrm{d}}(k)$.

Equation 4.16 is the final control solution of the standard DMC algorithm. For non constrained system, it can be directly implemented to calculate $\mathbf{U}(k)$. However, for systems with control constraints, the resulted control solution $\mathbf{U}(k)$ does not automatically fulfill the desired control constraints (such as equation 4.4). If the resulted $\mathbf{U}(k)$ does not fulfill the input constraint, an additional quadratic programming step is needed. Due to the modified system model 4.12, the form of input constraint also has to be adjusted as [Wan09]

$$\mathbf{U}_{\mathrm{d}}(k) = [\mathbf{I}_{\mathrm{d}}]\ \mathbf{U}(k-1) + [\boldsymbol{\Pi}_{\mathrm{d}}]\ \Delta\mathbf{U}_{\mathrm{d}}(k),$$

with

$$\mathbf{U}_{\mathrm{d}}(k) = \begin{bmatrix} \mathbf{U}(k) \\ \mathbf{U}(k+1) \\ \vdots \\ \mathbf{U}(k+p-1) \end{bmatrix}, \ \mathbf{I}_{\mathrm{d}} = \begin{bmatrix} [\mathbf{I}_M] \\ [\mathbf{I}_M] \\ \vdots \\ [\mathbf{I}_M] \end{bmatrix},$$

$$\mathbf{\Pi}_{\mathrm{d}} = \begin{bmatrix} [\mathbf{I}_M] & [\mathbf{O}_M] & \dots & [\mathbf{O}_M] \\ [\mathbf{I}_M] & [\mathbf{I}_M] & \dots & [\mathbf{O}_M] \\ \vdots & \vdots & \ddots & \vdots \\ [\mathbf{I}_M] & [\mathbf{I}_M] & \dots & [\mathbf{I}_M] \end{bmatrix},$$

where $[\mathbf{I}_M]$ and $[\mathbf{O}_M]$ are identical and zero matrices ($M \times M$), respectively. The dimensions of $[\mathbf{I}_{\mathrm{d}}]$ is $Mp \times M$ and the dimension of $[\mathbf{\Pi}_{\mathrm{d}}]$ is $Mp \times Mp$. Using the above notations, the control constraint in equation 4.4 is converted into

$$-[\mathbf{\Pi}_{\mathrm{d}}]^{-1}[\mathbf{I}_{\mathrm{d}}]\,\mathbf{U}(k-1) \leq \Delta\mathbf{U}_{\mathrm{d}}(k) \leq [\mathbf{\Pi}_{\mathrm{d}}]^{-1}\left(\mathbf{1} - [\mathbf{I}_{\mathrm{d}}]\,\mathbf{U}(k-1)\right), \tag{4.17}$$

where $\mathbf{1}$ is an all-ones vector ($Mp \times 1$) as

$$\mathbf{1} = [\,\underbrace{1, 1, 1, \dots, 1}_{Mp}\,]^{\mathrm{T}}.$$

The original control task (equation 4.15) can be described by a regular quadratic programming (QP) problem [GM86]

$$\begin{aligned} \min J_{\mathrm{LM}}(k) = \min \Big[& \Delta\mathbf{U}_{\mathrm{d}}^{\mathrm{T}}(k)\Big([\mathbf{\Xi}_{\mathrm{d}}(k)]^{\mathrm{T}}\,[\mathbf{\Gamma}]\,[\mathbf{\Xi}_{\mathrm{d}}(k)] + [\mathbf{\Lambda}] \Big)\Delta\mathbf{U}_{\mathrm{d}}(k) \\ & - 2\Delta\mathbf{U}_{\mathrm{d}}^{\mathrm{T}}(k)\,[\mathbf{\Xi}_{\mathrm{d}}(k)]^{\mathrm{T}}\,[\mathbf{\Gamma}]\,\Big(\mathbf{Y}_{\mathrm{d,t}}(k) - [\mathbf{F}_{\mathrm{d}}(k)]\,\mathbf{X}_{\mathrm{d}}(k)\Big)\Big) \\ & + \Big(\mathbf{Y}_{\mathrm{d,t}}(k) - [\mathbf{F}_{\mathrm{d}}(k)]\,\mathbf{X}_{\mathrm{d}}(k)\Big)^{\mathrm{T}}\mathbf{\Gamma}\Big(\mathbf{Y}_{\mathrm{d,t}}(k) - [\mathbf{F}_{\mathrm{d}}(k)]\,\mathbf{X}_{\mathrm{d}}(k)\Big)\Big], \end{aligned} \tag{4.18}$$

subject to

$$-[\mathbf{\Pi}_{\mathrm{d}}]^{-1}[\mathbf{I}_{\mathrm{d}}]\,\mathbf{U}(k-1) \leq \Delta\mathbf{U}_{\mathrm{d}}(k) \leq [\mathbf{\Pi}_{\mathrm{d}}]^{-1}\left(\mathbf{1} - [\mathbf{I}_{\mathrm{d}}]\,\mathbf{U}(k-1)\right),$$

The above QP problem can be solved using different numerical methods [Wan09]. In the practical MPC system of HEPHAISTOS, the control solution is firstly calculated using the standard DMC algorithm. If the resulted control vector $\mathbf{U}(k)$ is within the control constraint, it will be directly implemented. Otherwise, the control task will be transferred into the QP problem (equation 4.18) and solved using the numerical method.

Nonlinear MPC

There are generally two different ways to control the nonlinear system 3.46. The first one is to linearize the model and use linear MPC. According to the nonlinear model (equation 3.46), the microwave heating term is

$$\Psi^n(k) = \mathbf{V}^{\mathrm{T}}(k)\,[\mathbf{\Phi}^n(k)]\,\mathbf{V}(k).$$

The corresponding linearized term is

$$\begin{aligned}\Psi^n(k) &\approx \Psi^n(k-1) + \left.\frac{\partial \Psi^n}{\partial \mathbf{V}}\right|_{k-1} (\mathbf{V}(k) - \mathbf{V}(k-1)) \\ &\approx \Psi^n(k-1) + 2\mathbf{V}^{\mathrm{T}}(k-1)\,[\mathbf{\Phi}^n(k-1)]\,\big(\mathbf{V}(k) - \mathbf{V}(k-1)\big) \\ &\approx 2\mathbf{V}^{\mathrm{T}}(k-1)\,[\mathbf{\Phi}^n(k-1)]\,\mathbf{V}(k) - \mathbf{V}^{\mathrm{T}}(k-1)\,[\mathbf{\Phi}^n(k-1)]\,\mathbf{V}(k-1)\end{aligned} \tag{4.19}$$

which is a 'random walk' model [ZL03] depending on the former input vector $\mathbf{V}(k-1)$. In this case, errors from nonlinear system identification and linearization are accumulated, and this linearized model (equation 4.19) does not have a good prediction ability for future outputs. Therefore the linearization approach is not a good choice and the nonlinear MPC method has to be implemented.

Compared with the popularity of linear MPC, nonlinear MPC was not widely interested and studied until the 1990s [MHL99]. One of the most important driven reasons for the development of nonlinear MPC is the need for more accurate system models and better control performance. As mentioned previously, the control principle of nonlinear MPC is the same as linear MPC, except the prediction and control policy are derived from a nonlinear model. In our case, unlike the linear model, there is no analytical control solution derived from the cost function (equation 4.2) using the nonlinear model (equation 3.46). Therefore, a numerical control scheme is proposed combining the idea of bang-bang control [MTA+06] and genetic algorithms [Whi94], which is the binary genetic control scheme.

The so-called bang-bang control [MTA+06], also known as binary control or hysteresis control, is a control strategy where the control action switches between only two states [MTA+06]. When the bang-bang control scheme is used in the nonlinear model of HEPHAISTOS, it is defined that the control input variables switch between 0 and 1, which indicate the feeding source is turned off or turned on with full power respectively. Compared with normal control where the control variable can take any value between 0 and 1, it seems that the bang-bang control schemes loses a large part of the overall control diversity. But when the dimension of control inputs is large enough such as in HEPHAISTOS, the remained control diversity of such a binary control scheme is considerably large. For instance, in the new HEPHAISTOS Cavity 3, the number of feeding sources is 18 (the old HEPHAISTOS Cavity 3 had 36 sources), the total number of different heating combinations is 2^{18}. It means in theory the system could provide 2^{18} different heating patterns, which is still sufficient to provide a rather uniform heating.

To derive the control solution in the bang-bang control schemes, a binary programming algorithm [Sch98] has to be used. But due to the special nonlinear form of 3.46, normal binary programming algorithms are difficult to be implemented (most binary programming algorithms can only deal with linear optimization problems [Sch98]). Under this circumstance, the genetic algorithm (GA) is used here. GA is a powerful global optimization and search algorithm inspired by the natural evolution process [DAJ02]. Although many parts of GA are still not well understood such as its converging and stability properties, it can achieve surprisingly good performance for many practical problems where traditional algorithms could not work [HJK95]. Nowadays it has been one of the most successful optimization algorithms that are widely used in computer science, engineering, economics and other fields. Explicit introductions about GA can be found in [Whi94]. The binary GA based nonlinear MPC system of HEPHAISTOS contains the following four procedures.

- Step 1: Initialization

  At each time k, a number of possible control sequences are randomly generated, such as

$$\mathbf{V}_i^{\mathrm{c}} = \left[\mathbf{V}_i(k)^{\mathrm{T}} \ \mathbf{V}_i(k+1)^{\mathrm{T}} \ \ldots \ \mathbf{V}_i(k+p-1)^{\mathrm{T}} \right]^{\mathrm{T}}, \quad 1 \leq i \leq N_{\mathrm{g}},$$

where N_{g} is the total number of control sequences that are initialized. Each control sequence is called one individual or chromosome, and all individuals form a population. In the real application of HEPHAISTOS, it is assumed that

$$\mathbf{V}_i(k) = \mathbf{V}_i(k+1) = \cdots = \mathbf{V}_i(k+p-1).$$

That is because in MPC the first input vector $\mathbf{V}_i(k)$ takes more credits than the rest of the sequence. It may happen that a control sequence performs well referring to the cost function, but the first input vector is actually not a good choice. In order to avoid this situation, it is assumed that all following input vectors in the control sequence are identical to the first one, to fully examine the performance of $\mathbf{V}_i(k)$. Moreover, this arrangement also largely reduces the search space and raises the opportunity to quickly find the best control solution.

- Step 2: Evaluation and selection

  There are two important functions in GA. One is the evaluation function and the other one is the fitness function [Whi94]. Here the cost function (equation 4.2) is used as the evaluation function. All individuals of one population are substituted into the nonlinear model (equation 3.46) to predict their corresponding future temperature vectors, such as

$$\begin{aligned}
\mathbf{Y}_i(k+1) &= [\mathbf{A}(k)]\,\mathbf{Y}(k) + \mathbf{\Psi}\big(\mathbf{V}_i(k)\big),\\
\mathbf{Y}_i(k+2) &= [\mathbf{A}(k)]\,\mathbf{Y}(k+1) + \mathbf{\Psi}\big(\mathbf{V}_i(k+1)\big),\\
&\ \ \vdots\\
\mathbf{Y}_i(k+p) &= [\mathbf{A}(k)]\,\mathbf{Y}(k+p-1) + \mathbf{\Psi}\big(\mathbf{V}_i(k+p-1)\big).
\end{aligned}$$

  Above predicted temperature vectors are further substituted into the evaluation function 4.2 to calculate individual future costs J_i, such as

$$J_i = \sum_{l=k+1}^{k+p} \Big[\big(\mathbf{Y}_{\mathrm{t}}(l) - \mathbf{Y}_i(l)\big)^{\mathrm{T}} [\mathbf{\Gamma}(l)] \big(\mathbf{Y}_{\mathrm{t}}(l) - \mathbf{Y}_i(l)\big) + \mathbf{V}_i^{\mathrm{T}}(l)\, [\mathbf{\Lambda}(l)]\, \mathbf{V}_i(l) \Big],$$

  where $\mathbf{\Gamma}(l)$ and $\mathbf{\Lambda}(l)$ are defined the same as in equation 4.2.

The fitness function f_i is defined as

$$f_i = \exp(-\alpha \cdot J_i), \quad 1 \leq i \leq N_{\mathrm{g}},$$

where α is a tuning constant. The value f_i reflects how well the control sequence is fit for the optimization problem. After the fitness value is calculated for all individuals, the new generation of population will be selected. The principle of selection is defined as that for each old individual i, it has a probability of P_i to be selected as a new individual j in the new generation, where P_i is given by [Whi94]

$$\mathrm{P}_i = \frac{f_i}{\sum_{l=1}^{n} f_l}, \quad 1 \leq i \leq N_{\mathrm{g}}.$$

It is clear that the larger f_i is, the more suitable this control sequence is and the higher selected opportunity it has.

- Step 3: Crossover and mutation

  After the new generation is selected, the population randomly choose every two individuals into one pair and evolves these two individuals into new individuals by doing crossovers. In our case, the crossover occurs only for the first input vector and the rest of input vectors will replicate the first one, such as

$$\begin{array}{lll} \mathbf{V}_i^{\mathrm{c}} & \overbrace{\underline{0100}101011}^{\mathbf{V}_i(k)}\ \overbrace{\underline{0100}101011}^{\mathbf{V}_i(k+1)}\ldots & \Longrightarrow \quad \underline{1100}101011\ \underline{1100}101011\ldots \\ \mathbf{V}_j^{\mathrm{c}} & 10101\underline{11001}\ 10101\underline{11001}\ldots & \quad 10101\underline{01001}\ 10101\underline{01001}\ldots \end{array}$$

  The length of crossover is predefined, but the starting point of crossover could be determined randomly.

  After crossover, the next step is mutation. For each bit of $\mathbf{V}_i^{\mathrm{c}}$, it has a tiny probability P_m to be switched to the other state (from 0 to 1 or from 1 to 0). The probability is defined as $\mathrm{P}_m = 1/(mp)$, where mp is the number of total bits in each individual. Similarly with the crossover part, mutation is also taken only for bits of the first input vector.

- Step 4: Termination

  The new individuals generated from crossover and mutation will be evaluated and selected again to create another new generation. Steps 2 and 3 are iterated until the number of generations is reached or a certain level of fitness value is achieved. The control sequence with the highest fitness value will be selected, and the first control input vector within this sequence will be applied as the real control input vector $\mathbf{V}(k)$.

The biggest advantage of GA is that the whole process can search through the whole input space efficiently without being trapped in a local optimum [Whi94]. During the searching process, a large number of predictions have to be performed, therefore the time spent on each single prediction is a key element that affects the performance of the whole algorithm. In order to have more iterations within a given control period, the time for each prediction should be as short as possible. Compared with multiple layers calculations using nonlinear activation functions in the neural network estimator, the time required by the nonlinear model (equation 3.46) is much less. Therefore, the nonlinear model is used for the prediction and GA controlling.

Stability is always one of the most important and fundamental issues in the controller design. For the linear MPC, the general stability of DMC is proved in [GM82] in situations where the prediction horizon length p is significantly larger than the dimension of the input vector $p >> n$. It was further demonstrated in [Cut83] that DMC is stable in the case of $p \geq n + n_\mathrm{d}$ where n_d is the largest input to output delay time. So far there is no explicit proof about that under what circumstances a finite horizon input constrained QDMC algorithm is closed-loop stable. However, as mentioned in [GM86] that in general cases QDMC is more stable and robust than DMC, because of the loss of control gain due to input constraints [Mor85]. The stability of QDMC has also been verified by a large number of successful applications in the oil industry.

Compared with the QDMC algorithm, the stability conditions of nonlinear MPC has not been well established. To the author's knowledge, currently there are no practically useful stability results established for GA. Nevertheless, a number of empirical approaches can be done to improve their stabilities and robustnesses in practical applications.

For example, according to empirical evidences from numerous experiments of HEPHAISTOS, a good upper power limit U_{max} can be obtained by using a simple proportional-integral (PI) controller. In this case, the number of totally activated feeding sources in the GA controller is always bounded by the power calculated by the PI controller, which not only guarantees the control stability but also reduces the searching range of the input space. The practical GA based nonlinear MPC system is shown as figure 4.3.

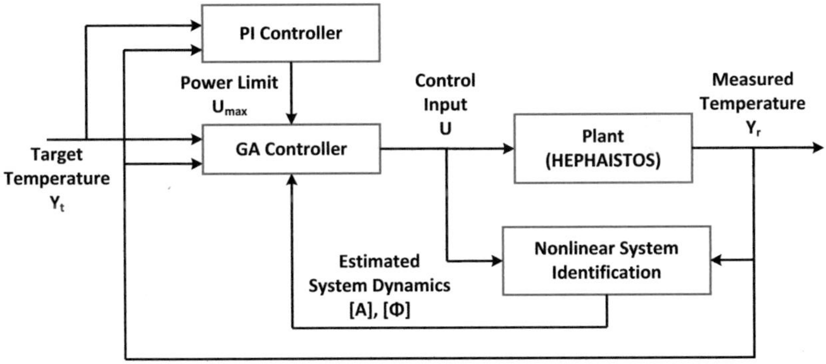

Figure 4.3. Practically implemented GA based nonlinear MPC system.

4.1.2. Neural Network based Control

The idea of using neural networks as the controller has been studied and implemented massively in numerous applications [PSRJG00] [LP02] [PW08]. In general there are two main structures that apply neural network based control (NNC) [LV09], such as in figure 4.4.

The first structure (4.4a) is called indirect NNC, which has a similar topology with traditional feedback control system and involves both the NN estimator and the NN controller. The estimator learns the dynamics of the unknown system and the controller uses the estimated model to control the real plant. The second structure (4.4b) contains only a NN controller which directly controls the real plant.

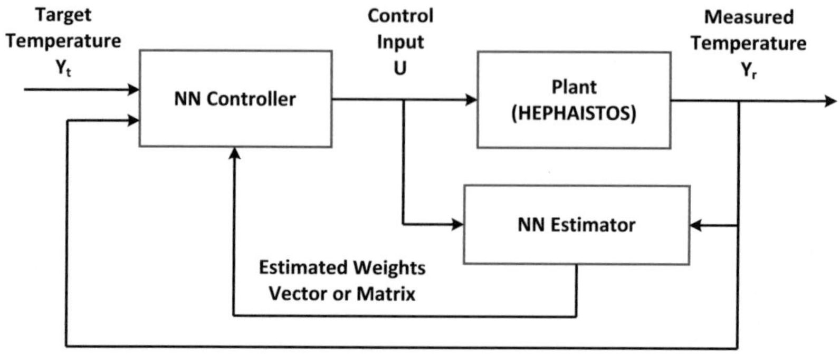

(a) Indirect control structure

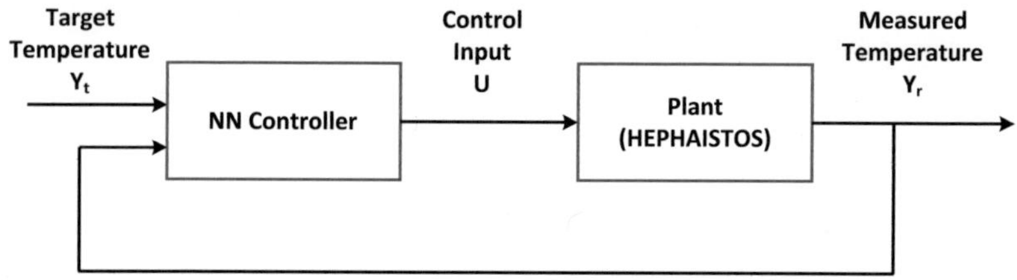

(b) Direct control structure

Figure 4.4. Two control structures of NN controller.

The weights of this NN controller is usually tuned using the unsupervised learning algorithms. Unlike in supervised learning, where the target output of the NN is given in advance, in unsupervised learning there is no explicit target output provided. Therefore the main challenge in using a direct NN controller is to determine a suitable and realistic learning algorithm that can effectively tune the weights and guarantee the control performance.

The dynamics of the real plant to be controlled can be described by the expression, such as

$$\mathbf{Y}(k+1) = F\left(\mathbf{Y}(k), \mathbf{U}(k)\right), \tag{4.20}$$

and the neural network controller is represented by

$$\begin{aligned} \mathbf{U}(k) &= G\left(\mathbf{Y}_{\mathrm{t}}(k), \mathbf{Y}(k), \mathbf{w}(k)\right), \\ \mathbf{w}(k) &= \left[w_1(k),\ w_2(k),\ \ldots,\ w_{\mathrm{N_w}}(k)\right]^{\mathrm{T}} \end{aligned} \tag{4.21}$$

where F is the function that describes the dynamics of the real plant, G is the function of the NN controller. and $\mathbf{w}(k)$ is the weight vector of the NN controller ($\mathrm{N_w} \times 1$). Equation 4.21 means the control action is directly calculated based on the target and current measured temperatures as well as the weights of the NN. With the same cost function in equation 4.2, the objective of unsupervised learning in this NNC is defined to find the optimal weight vector $\mathbf{w}^*$ that minimizes the cost function, which is

$$\mathbf{w}^* = \arg_{\mathbf{w}} \min J(k) \quad \Rightarrow \quad \frac{\partial J(k)}{\partial \mathbf{w}^*} = 0. \tag{4.22}$$

The weight update in the NN controller still uses the standard gradient descent principle, following the direction that minimizes the cost function, such as

$$\begin{aligned} \mathbf{w}(k+1) &= \mathbf{w}(k) - \alpha(k) \cdot \left.\frac{\partial J(k)}{\partial \mathbf{w}}\right|_k \\ &= \mathbf{w}(k) - \alpha(k) \cdot \left.\frac{\partial J(k)}{\partial \mathbf{Y}} \cdot \frac{\partial \mathbf{Y}}{\partial \mathbf{U}} \cdot \frac{\partial \mathbf{U}}{\partial \mathbf{w}}\right|_k, \end{aligned} \tag{4.23}$$

where $\alpha(k)$ is the step size of the update.

If the dynamics of the real plant F is perfectly known, the partial differentiation $\partial\mathbf{Y}/\partial\mathbf{U}$ can be calculated and the gradient term can be directly implemented. However, for most cases, including HEPHAISTOS, the real dynamics of the controlled plant is incompletely known, and the partial differential term $\partial\mathbf{Y}/\partial\mathbf{U}$ is not directly computable. In this case, a stochastic approximation (SA) algorithm [KY97] has to be applied [SC98]

$$\mathbf{w}(k+1) = \mathbf{w}(k) - \alpha(k) \cdot (\,\text{approximated gradient}\,)_k\,, \tag{4.24}$$

which replaces the true gradient by an approximated gradient to update the weights of the NN controller.

SA is a powerful method that is used for optimization problems where traditional analytical methods could not solve. In this section, the so-called simultaneous perturbation stochastic approximation (SPSA) learning algorithm is introduced to demonstrate how the NN controller is designed and applied in HEPHAISTOS. SPSA was firstly developed in [Spa87] as a general SA algorithm used in parameter estimation. Since the middle of the 1990s, it has been gradually extended to the field of neural network learning and control, and widely implemented in practical applications, such as in [SC94] [Spa98] [BK04] [SSSN08]. The procedures to implement SPSA in a NN controller are described in the following.

At each time k, the weight vector of the NN controller is updated by the equation

$$\mathbf{w}(k+1) = \mathbf{w}(k) - \alpha(k) \cdot h\left(\mathbf{w}(k)\right), \tag{4.25}$$

where $h\left(\mathbf{w}(k)\right)$ is the simultaneous perturbation approximation of the original gradient $\partial J(k)/\partial \mathbf{w}$. The approximation term is calculated by the equation

$$h\left(\mathbf{w}(k)\right) = \frac{J^{(+)}(k) - J^{(-)}(k)}{2\, c(k)\, \boldsymbol{\Delta}(k)}, \tag{4.26}$$

where

$$\begin{aligned}
\boldsymbol{\Delta}(k) &= \left[\,\Delta_1(k), \Delta_2(k), \ldots, \Delta_{\mathrm{N_w}}(k)\,\right]^{\mathrm{T}}, \\
\mathbf{w}^{(\pm)}(k) &= \mathbf{w}(k) \pm c(k)\, \boldsymbol{\Delta}(k), \\
\mathbf{U}^{(\pm)}(k) &= G\left(\mathbf{Y}_{\mathrm{t}}(k), \mathbf{Y}(k), \mathbf{w}^{(\pm)}(k)\right), \\
\mathbf{Y}^{(\pm)}(k) &= \mathbf{Y}(k+1) = f\left(\mathbf{Y}(k), \mathbf{U}^{(\pm)}(k)\right), \\
J^{(\pm)}(k) &= J\left(\mathbf{Y}^{(\pm)}(k), \mathbf{U}^{(\pm)}(k)\right)\Big|_{p=1},
\end{aligned} \tag{4.27}$$

and $c(k)$ is the tuning constant that fulfills certain regularity conditions. Several points need to be noticed regarding the implementation of the learning process:

- The simultaneous perturbation vector $\boldsymbol{\Delta}(k)$ is randomly generated. All elements $\Delta_i(k)$ are independent, bounded and symmetrically distributed random variables [SC98]. In practice, these elements could be randomly generated around 0 with amplitudes smaller than 1. Another simpler method is to randomly choose $\Delta_i(k)$ to be either 1 or -1. Different methods of determining $\Delta_i(k)$ can be applied accordingly.

- In the standard SPSA control system, one complete weight update of $\mathbf{w}(k)$ takes three control periods. In the beginning time k, the NN controller generates $\mathbf{\Delta}(k), \mathbf{w}^{(\pm)}(k)$ and $\mathbf{U}^{(\pm)}(k)$. and implements the new control action $\mathbf{U}^{(+)}(k)$. At time $k+1$, it measures corresponding system output $\mathbf{Y}^{(+)}(k)$ (which is also $\mathbf{Y}(k+1)$), calculates the first evaluation value $J^{(+)}(k)$ using $\mathbf{Y}^{(+)}(k)$, and then implements $\mathbf{U}^{(-)}(k)$. The same procedures are repeated at time $k+2$ to calculate $J^{(-)}(k)$. After that the weight $\mathbf{w}(k)$ will be updated using equation 4.25. The resulted new weight is not strictly $\mathbf{w}(k+1)$, but $\mathbf{w}(k+3)$ which will be used to calculate new control action for the period $k+3$. In principle, the output sequence can be expressed as (see figure 4.5)

$$O = \{\ldots, \mathbf{Y}(k), \mathbf{Y}^{(+)}(k), \mathbf{Y}^{(-)}(k), \\ \mathbf{Y}(k+3), \mathbf{Y}^{(+)}(k+3), \mathbf{Y}^{(-)}(k+3), \ldots\}. \tag{4.28}$$

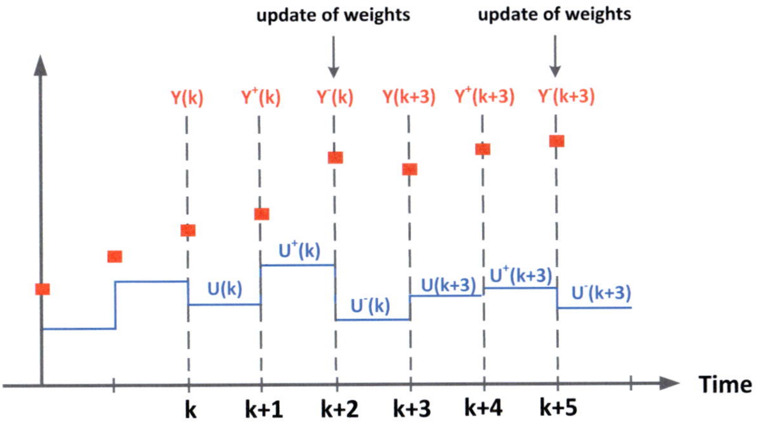

Figure 4.5. Weights update in the standard SPSA algorithm.

The standard SPSA algorithm (equation 4.26) has a critical limitation that makes it difficult to be realized in practice. It is defined in equation 4.27 that both $\mathbf{Y}^{(+)}(k)$ and $\mathbf{Y}^{(-)}(k)$ are generated based on $\mathbf{Y}(k)$, which means in the sequence (equation 4.28) the controller should wait until $\mathbf{Y}^{(+)}(k)(\mathbf{Y}(k+1))$ equivalent to $\mathbf{Y}(k)$ to implement the next control action $\mathbf{U}^{(-)}(k)$. This condition requires extra waiting time and also brings more disturbances for continuous multiple-output systems, where two adjacent outputs are hardly identical. In addition, the weight update frequency in the standard SPSA algorithm is limited as one update per three control periods, which slows down the whole controller converging speed and correspondingly degrades the control performance.

In order to make the SPSA algorithm more efficient in the practical heating process, a semi-direct NN controller is implemented in this dissertation. The first modification is to change the approximation of the gradient from equation 4.26 to a simpler form as [SC98]

$$h(\mathbf{w}(k)) = \frac{J^{(+)}(k)}{c(k)\,\mathbf{\Delta}(k)}. \tag{4.29}$$

Although the two-measurement strategy (equation 4.26) is generally more preferable, it has been proved in [Spa97] that this one-measurement strategy (equation 4.29) is also suitable for highly non-stationary systems, where the parameters of the plant or external disturbances might change during one control period (from $k+1$ to $k+2$ or from $J^{(+)}(k)$ to $J^{(-)}(k)$). In order to further speed up the learning process, the control structure is also modified from the direct control scheme to a semi-direct scheme that involves an additional NN estimator (figure 4.6).

In this semi-direct scheme, the NN estimator keeps learning the dynamics of the real plant and monitoring the error between its prediction and real output of the plant. If the prediction of the NN estimator is constantly accurate for certain time (the MSE of prediction is lower than a given threshold, described in figure 4.7), it will also be used in the weight update. This NN estimator can be considered as an approximation of the real plant, to provide information such as $\hat{\mathbf{Y}}^{(+)}(k), \hat{J}^{(+)}(k)$ to the NN controller. During each control period, the NN controller can take one or more updates based on the NN estimator, and the number of updates per control period can be adjusted. The principle can be found in the controller description (figure 4.7).

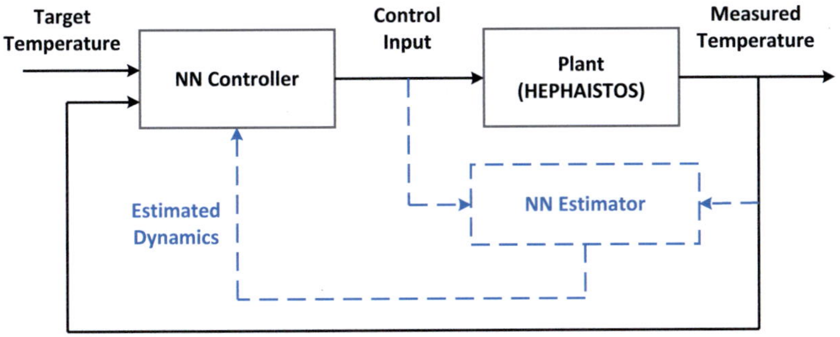

Figure 4.6. Semi-direct control structure. The blue dashed line represents two possible learning approaches.

The main advantage of the semi-direct control structure is that the overall weight update of the NN controller is significantly accelerated. The average weight update rate can be increased to once per control period or even higher. In addition, there is no extra waiting time spent on the states or outputs recovery. The NN controller using one-measurement SPSA learning can quickly evolve from the initialized state to a nearly optimal state, which guarantees acceptable control performance in the starting period. Defining the modulo operation as

$$r = \mathrm{Mod}(a, n) = a - n\lfloor\frac{a}{n}\rfloor, \quad \text{for} \quad a, n, r \in R,$$

where r is the reminder of a divided by n.

The complete procedures of the semi-direct control scheme is shown in figure 4.7.

4.2. Intelligent Control

Intelligent control is a class of control methods that utilize ideas and approaches from the artificial intelligence or biological systems to solve the control problem [WSF94]. Technically, both aforementioned nonlinear GA control and NNC methods can be classified as intelligent control approaches. Nevertheless, in this dissertation intelligent control is defined not only by characterizing applied control methods, but also by characterizing the control task being optimized. In other words, the control task used in the intelligent control system is also "intelligent". The so-called intelligent control task is named in comparison with traditional control tasks. In conventional heating problems, the mean square error (MSE) between measured temperatures and the target temperature is used as the cost (objective) function of the control task, such as equation 4.2 used in the MPC method.

In many cases, the expression 4.2 is not always the best choice for the cost function. On the one hand, the true temperature distribution is not well represented by the value of the cost function. For example, during the heating process, there are a number of hot spots and cold spots that are not measured, and thus not reflected by the value of the cost function (equation 4.2). On the other hand, in applications where the entire temperature profile is available, only a small part of the profile is used in the control of the cost function 4.2. The most part of the information from the thermal picture is wasted.

Initialize weights of the NN controller and the NN estimator as $\mathbf{w}_{\mathrm{c}}$ and $\mathbf{w}_{\mathrm{e}}$;

Initialize functions of the NN controller and the NN estimator as G and F;

Set the estimation error indicator χ_{ind} and its lower threshold χ_{th};

Set the parameters $c(k), \alpha(k)$, the counter number $Ct = 1$ and the indirect training timer $n_{\mathrm{t}} = 1$;

Take the first control action as $\mathbf{U}(k) = F_{\mathrm{c}}(\mathbf{Y}(k), Y_{\mathrm{tar}}(k))$ at time $k = 1$;

Figure 4.7. Procedures in the semi-direct NN control system (part 1).

In this case, it is better to replace the traditional cost function by more innovative and reliable definitions, which are the intelligent control tasks as previously mentioned. For example, using a infrared camera to measure the temperature profile of the heated load, the maximum and the minimum temperatures of the whole heated load $Y_{\mathrm{max}}, Y_{\mathrm{min}}$ can be obtained in real time, and a more reliable cost function can be

At each time k $(k > 1)$:

repeat

Receive the measured temperature vector $\mathbf{Y}(k)$;

Calculate the estimation RMSE of the NN estimator as $\chi(k)$, and update the estimation error index via

$$\chi_{\mathrm{ind}} = 0.95 \cdot \chi_{\mathrm{ind}} + 0.05 \cdot \chi(k).$$

Update the weights of the NN estimator using the EKF algorithm;

if $\mathrm{Mod}(Ct, 2) = 0$ **then**

  if $\chi_{\mathrm{ind}} < \chi_{\mathrm{th}}$ **then**

    Use the NN estimator to train the NN controller for n_{t} times to get new $\mathbf{w}_{\mathrm{c}}(k)$ and update the training timer $n_{\mathrm{t}} = n_{\mathrm{t}} + 1$;

    Calculate the new control input $\mathbf{U}(k) = G\left(Y_{\mathrm{tar}}(k), \mathbf{Y}(k), \mathbf{w}_{\mathrm{c}}(k)\right)$;

  else

    Reset the training timer $n_{\mathrm{t}} = 1$;

    Calculate the training control input as

$$\mathbf{w}^{+}(k) = \mathbf{w}(k) + c(k)\,\boldsymbol{\Delta}(k),$$
$$\mathbf{U}(k) = \mathbf{U}^{+}(k) = G\left(\mathbf{Y}_{\mathrm{t}}(k), \mathbf{Y}(k), \mathbf{w}^{(+)}(k)\right);$$

    Update the counter number $Ct = Ct + 1$;

else

  Update the weights of the NN controller by

$$h\left(\mathbf{w}_{\mathrm{c}}(k)\right) = \frac{J\left(\mathbf{Y}(k), \mathbf{U}^{(+)}(k)\right)}{c(k)\,\boldsymbol{\Delta}(k)}$$
$$\mathbf{w}_{\mathrm{c}}(k+1) = \mathbf{w}_{\mathrm{c}}(k) - \alpha(k) \cdot h\left(\mathbf{w}_{\mathrm{c}}(k)\right);$$

  Update the counter number $Ct = Ct + 1$;

  Calculate the new control input $\mathbf{U}(k) = G\left(Y_{\mathrm{tar}}(k), \mathbf{Y}(k), \mathbf{w}_{\mathrm{c}}(k)\right)$;

until *the end of the control process;*

Figure 4.7. Procedures in the semi-direct NN control system (part 2).

generated as

$$J_{IC} = \frac{1}{N} \sum_{i=1}^{N} (Y_i - Y_t)^2 + \alpha_{\max} (Y_{\max} - Y_t)^2 + \alpha_{\min} (Y_{\min} - Y_t)^2 , \tag{4.30}$$

where $\alpha_{\max}$ and $\alpha_{\min}$ are coefficients to adjust weights of $Y_{\max}$ and $Y_{\min}$, respectively. In equation 4.30, the maximum and the minimum temperatures of the whole load are also included in the cost function, which is more accurate to reflect the true temperature homogeneity of the heated load. However, the problem regarding this cost function is that the maximum and the minimum temperatures are not fixed in certain locations, and their locations are varying during the heating process. It is not feasible to build a model to describe the relationship between the maximum and the minimum temperatures and individual microwave feeding power. As a result, normal control techniques are difficult to be applied in this situation.

4.2.1. Reinforcement Learning

The development and implementation of reinforcement learning (RL) in the control field provides a perfect alternative to deal with aforementioned intelligent control tasks. The idea of reinforcement learning was originally inspired by the biological learning process [LV09], like the learning behaviors of human beings and other animals. In RL if an action is followed by a satisfactory state of affairs or an improvement in the state of affairs, it will receive a positive (or less negative) reward and the tendency to produce that action is strengthened, which is reinforced. Otherwise if an action is followed a non-satisfactory state of affairs, it will receive a negative (worse than the first case) reward and the tendency to produce that action is weakened [Bar94]. The ultimate objective of RL is to find the optimal action (control) policy that maximizes the overall rewards obtained during the entire control process.

Unlike the supervised learning approach introduced in the previous chapter 3, where the learning is based on datasets provided by a knowledgable supervisor, there is no predefined learning target given in RL. In most cases the plant to be controlled is completely unknown to the learner. Therefore the learner has to use a trial-and-error searching strategy to consistently explore the environment by performing different actions, and improve the action policy according to the received rewards from the environment. RL is more practical and useful than supervised learning in many cases, because it is often impossible to obtain training datasets that are both correct and representative to describe all dynamics of the real problem. That is also why RL is considered as the closest learning approach to functionalities of the human brain, and widely applied and studied in artificial intelligence, robotics and computer science areas [GBLB12].

RL was extended and implemented in the control field starting from the 1980s and 1990s [BSB81] [Sut84], and gradually developed as an influential control method [SBW92] [WD92] [WMS92]. The main difference between RL control (RLC) and other control methods is that in RLC a different tool is utilized to describe the dynamics of the plant being controlled. Instead of transfer equations (polynomials) or state-space models used in conventional control methods, in RL the plant is modeled by a Markov decision process (MDP) [Bel57] that only consists of different state-action pairs and transition probabilities between any two states. The complete dynamics of the plant are described by the probability distribution

$$P_a(s, s', r) = \Pr\left(R(k+1) = r, S(k+1) = s' \mid S(k) = s, A(k) = a\right), \tag{4.31}$$

where $P_a(s, s', r)$ is the transition probability from state s to state s' with a reward r, caused by the action a at time k.

From a conventional control engineering point of view, the action $A(k)$ is equivalent to the control input ($\mathbf{U}(k)$) decided by the controller, and the state $S(k)$ is the basis for making control actions, which functions similarly with the output variable ($\mathbf{Y}(k)$). The reward $R(k)$ is the basis for evaluating the control actions, to tell a control action is good or not, and it has the same functionality with the cost defined by the cost function (such as equation 4.30). In order to keep a good consistency in matters of notations and descriptions, in the following the symbols $\mathbf{U}(k)$ and $\mathbf{u}$ are utilized to replace $A(k)$ and a, as the selected control action at time k and the random control action (vector), respectively.

Meanwhile, the name 'plant' is used to denote the external environment in RL.

In most cases, states in a MDP are discrete and finite, but the above expression 4.31 can be extended into situations of infinite or continuous states [SK00]. An important property always assumed in MDPs is that the next state $S(k+1)$ and reward $R(k+1)$ depends only on the current state $S(k)$ and action $\mathbf{U}(k)$, which makes the MDP similar to an ARX model with one-step delay. For plants where the above property does not fully hold, it is still appropriate to consider them as approximated MDPs, as long as the current state can provide a good basis for predicting the next state and reward.

Despite significant differences regarding the process of system modeling and the controller design, it was found that RLC is closely related to conventional control methods. It has been proved in [SBW92] that RLC is essentially an optimal control approach, and the relationship between RLC and adaptive feedback control was well explained in [LV09]. More and more applications implement the principle of RL in conventional control frameworks, such as the RL based fuzzy controller [JLL00] [Lin03] and RL based online PID tuning algorithm [HB00]. Detailed information of RL refers to literatures as [Bar98], [Alp04] and [B$^+$06].

The structure of a normal RLC system is shown in figure 4.8. In general, a RLC system consists of following four parts.

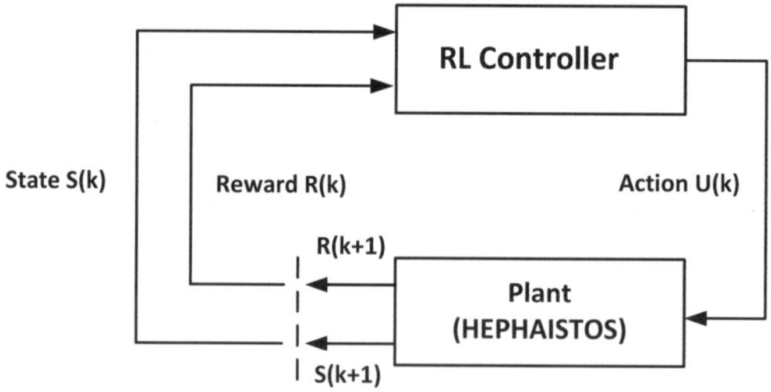

Figure 4.8. Reinforcement learning controller.

Control policy $\pi(s, \mathbf{u})$

The control policy is a stochastic or deterministic rule by which the controller decides its future (next) actions based on the current state. It could be represented by a lookup table, a probability distribution or a function, mapping from individual states to control actions to be taken under each state. The control policy is the core of a RLC, which is equivalent to the control solution in conventional controllers.

Reward function $R(k)$

The reward function is defined as the future rewards that the controller tries to maximize. It maps each control action to a scalar value, which is the reward, evaluating the intermediate desirability of the action. If the plant goes from a state of less (higher) value (the value of a state is determined by the value function such as introduced below) to a state of higher (less) value, the corresponding reward is positive (negative), indicating this transition is good (bad) for the plant and controller. The value of rewards could be finite predefined values or calculated results from a function, depending on different formulations of the reward function.

Value function

A value function specifies the long-term expected rewards for each state or state-action pair under a certain control policy. There are two types of value function defined. One is the state value function $V_\pi(s)$, which denotes the long-term expected reward starting from the state s using the control policy π. The state value function is defined as (according to [Bar98])

$$V_\pi(s) = E\left[\,G(k) \mid S(k) = s, \pi\,\right], \tag{4.32}$$

where $G(k)$ represents the long-term expected reward. The long-term expected reward can be formulated according to different criteria, and two mostly used formulations are the averaged long-term expected reward [Bar98]

$$G(k) = \lim_{n\to\infty} \frac{1}{n} \sum_{k=1}^{n} R(k), \tag{4.33}$$

and the discounted long-term expected reward [Bar98]

$$G(k) = \sum_{k=1}^{n} \gamma^{k-1} R(k), \tag{4.34}$$

where $0 < \gamma < 1$ is the discount factor. The involvement of the discount factor is to avoid situations where the values of certain states go to infinite during the learning process. In this dissertation, the discounted long-term expected reward (equation 4.34) is used to within value functions.

The other type of value function is called state-action value function $Q_\pi(s, \mathbf{u})$ (Q-function), which is defined as the expected long-term reward starting from the state s, taking the action a and thereafter following the control policy π. It also has different formulations such as the above shown state value function. Here the expression similar to equation 4.34 is used as the state-action value function [Bar98]

$$\begin{aligned} Q_\pi(s, \mathbf{u}) =& E\left[\, G(k) \,|\, S(k) = s, \mathbf{U}(k) = \mathbf{u}, \pi \,\right] \\ =& E\left[\sum_{k=1}^{n} \gamma^{k-1} R(k) \,\middle|\, S(k) = s, \mathbf{U}(k) = \mathbf{u}, \pi \right]. \end{aligned} \tag{4.35}$$

Whereas rewards determine the immediate, intrinsic desirability of individual states of the plant, the value function indicate the long-term desirability of states or state-action pairs by taking into account succeeded states and rewards [Bar98].

The relationship between the state and state-action value functions are represented by the following two equations [Bar98]

$$V_\pi(s) = \sum_{\mathbf{u}} \pi(\mathbf{u}|s) Q_\pi(s, \mathbf{u}), \tag{4.36}$$

$$Q_\pi(s, \mathbf{u}) = \sum_{s'} P_u(s, s', r) \left[r(s, \mathbf{u}, s') + \gamma V_\pi(s') \right], \tag{4.37}$$

where $\pi(\mathbf{u}|s)$ is the probability of taking action u in the state s under the control policy π. Combining these two equations together, there is the so-called Bellman equation given as [Bar98]

$$V_\pi(s) = \sum_{\mathbf{u}} \pi(\mathbf{u}|s) \sum_{s'} P_u(s, s', r) \left[r(s, \mathbf{u}, s') + \gamma V_\pi(s') \right]. \tag{4.38}$$

Defining the optimal state and state-action value functions as

$$V^*(s) = \max_\pi V_\pi(s), \tag{4.39}$$

$$Q^*(s, \mathbf{u}) = \max_\pi Q_\pi(s, \mathbf{u}), \tag{4.40}$$

respectively, the Bellman optimality equations are expressed as [Bar98]

$$\begin{aligned} V^*(s) &= \max_{\mathbf{u}} \mathbf{E} \left[R(k+1) + \gamma V^*(S(k+1)) \,\middle|\, S(k) = s, \mathbf{U}(k) = \mathbf{u} \right] \\ &= \max_u \sum_{\mathbf{u}} \pi(\mathbf{u}|s) \sum_{s'} P_u(s, s', r) \left[r(s, \mathbf{u}, s') + \gamma V^*(s') \right], \end{aligned} \tag{4.41}$$

$$\begin{aligned} Q^*(s, \mathbf{u}) &= E \left[R(k+1) + \gamma \max_{\mathbf{u}'} Q^*(s, \mathbf{u}') \,\middle|\, S(k) = s, \mathbf{U}(k) = \mathbf{u} \right] \\ &= \sum_{s'} P_u(s, s', r) \left[r(s, \mathbf{u}, s') + \gamma \max_{\mathbf{u}'} Q^*(s', \mathbf{u}') \right]. \end{aligned} \tag{4.42}$$

The Bellman optimality equation is the foundation of RLC as well as the conventional optimal control [LV09]. It reveals the relationship between the current state (or state-action pair) and its successor state (or state-action pair), which transfers the process of determining the long-term optimal control sequence into a one-step search of equation 4.41

or 4.42 . In other words, any control policy that is greedy [ZSWM00] [RS07] with respect to the optimal state-action value function 4.42 is an optimal policy [Bar98], such as

$$\pi^*(s, \mathbf{u}) : \quad \mathbf{u} = \mathbf{u}^* = \arg\max_{\mathbf{u}} Q^*(s, \mathbf{u}), \tag{4.43}$$

where $\mathbf{u}^*$ indicates the optimal control action.

Model of the plant

The function of the model is to mimic the dynamics of the plant and predict future states and rewards. It is not a compulsory element for all RLC methods. Classical RLC methods use the pure trial-and-error strategy, where all of their learning and approximations are based on explicitly experienced interactions, and the model of the plant is not needed. In modern RLC methods, planning based on the model of the plant is often utilized to speed up the learning process [Die99]. Besides real experiences, simulated experiences from the model of the plant is also useful to improve and update the control policy [AR01]. Although RLC with the incremental/online planning costs more computation power than the direct RLC, it showed in [AS97] and [LV09] that a faster learning speed and a higher expected return can be obtained because of the involvement of explicit models.

Several points are worthy of note regarding these four parts. Firstly, RL is able to deal with different types of problems and control tasks, as long as the value function is clearly defined and rewards are appropriately assigned. From this point of view, the task (equation 4.30) can be solved using RL. Secondly, In practical control applications, the state-action value function 4.35 makes more sense than the state value function 4.34. Because as shown in equation 4.43, the state-action value function directly defines how good or bad a control action is, and the optimal controller can be simply constructed using the greedy algorithm, which makes the controller design more straightforward than using the state value function. Due to this reason, all following RL methods are introduced in the the state-action value function form. Finally, as previously emphasized, the value function is the core of all RLC methods. Using different approaches to calculate or approximate the value function and then using the value function to derive the control policy are the procedures followed by all RLC methods.

Temporal difference methods

According to different applicable situations and different approaches of updating the value function, RLC can be classified into three types: dynamic programming (DP) [BBBB95] [Si04], Monte Carlo (MC) [BC06] methods and temporal difference (TD) [Tes92] methods. DP is the most basic and classical RLC scheme, which requires complete knowledge of the plant. When the complete dynamics of the plant are known in advance, the transition probability distribution is easily obtained. The value function and corresponding optimal control policy can be directly calculated via iterative updates according to equations 4.37 and 4.43, respectively. In DP methods, both the value function and the control policy are updated simultaneously through interactions between each other, using either the policy iteration (PI) algorithm, the value iteration (VI) algorithm and other generalized policy iteration (GPI) algorithms. Differences between among algorithms and detailed introductions can be found in [KLM96] and [Si04]. DP methods are guaranteed to converge to the final optimality for finite MDPs, but they are thought of limited usage in practice because of its inefficiency for high-dimensional problems and requirement of complete knowledge of the environment [Bar98].

When the dynamics of the plant are not completely known, both MC and TD methods can be applied to approximate the value function based on real time experiences. MC is a learning strategy based on random explorations. In MC methods, the update of the value function is done episode by episode [BC06]. Each episode starts from a random state, takes control actions defined randomly or from a control policy π, and ends at a predefined terminal state. For each state occurred in an episode, its value function is simply calculated as the accumulated rewards from its first (or every) appearance until the end of the episode. After multiple episodes, the mean value function can be obtained and the greedy algorithms is applied to generate the final deterministic control policy (such as equation 4.43). The idea of MC methods is easy but the main problem is that the update is only performed at the end of each episode rather than at each time step, which makes MC methods inconvenient to be implemented in the online form.

TD methods are the combination of DP and MC [Tes95]. On the one hand, TD methods can estimate the the value function directly from raw experiences without any prior information of the plant, just like MC methods. In addition, the update of the value function in TD is easy to be implemented in the online form without waiting until the end of each episode, which is similar to DP methods. As the name indicates, the update of the state-action value function in TD methods is based on the so-called TD error [Tes95]. Depending on different definitions of TD error and different update rules, TD methods can be classified into three types as following [Bar98].

- Sarsa (State-Action-Reward-State-Action)

  At each time k, the update rule for the state-action value function in sarsa is defined as [Bar98]

$$\begin{aligned} Q(S(k), \mathbf{U}(k)) =& Q(S(k), \mathbf{U}(k)) + \alpha(k)\,[\,R(k) \\ &+ \gamma Q(S(k+1), \mathbf{U}(k+1)) - Q(S(k), \mathbf{U}(k))\,], \end{aligned} \tag{4.44}$$

  where $\alpha(k)$ is the time-varying learning rate and the term within the square brackets is used as the TD error as [Bar98]

$$\delta_{\mathrm{td}} = R(k) + \gamma Q(S(k+1), \mathbf{U}(k+1)) - Q(S(k), \mathbf{U}(k)). \tag{4.45}$$

  In order to guarantee the state-action value function converges to the optimal value function, the time-varying learning rates have to fulfill the condition as [Bar98]

$$\sum_{k=1}^{\infty} \alpha(k) = \infty, \quad \text{and} \quad \sum_{k=1}^{\infty} \alpha(k)^2 < \infty. \tag{4.46}$$

  After the state-action value function is updated, the controller can be constructed by using the greedy algorithm (equation 4.43), or the more robust ϵ-greedy algorithm, such as [Bar98]

$$\pi(\mathbf{u}|s) = \begin{cases} 1 - \epsilon + \epsilon/|U(s)|, & \mathbf{u} = \mathbf{u}^*, \\ \epsilon/|U(s)|, & \text{others}, \end{cases} \tag{4.47}$$

where $|U(s)|$ is defined as the total number of available control actions at the state s and $0 \leq \epsilon \leq 1$ is the probability factor representing the total probability to select other non-greedy control actions.

The involvement of the probability factor ϵ is to achieve a trade-off between exploration and exploitation, which is important for not only sarsa but all TD methods [Thr92] [AN05] [ALL$^+$09]. On the one hand, the controller has to make the best action according to the greedy algorithm (exploitation), trying to obtain a high long-term expected reward. But on the other hand, the controller should also occasionally explore other non-greedy actions to test if there are better control actions. This is helpful for future improvements of both the value function and the control policy, especially when the value function is not optimal, or when the plant is stochastic or time-varying.

The value of the probability factor ϵ is determined mainly depending on the plant. In principle, the value of ϵ is large in the beginning of the learning process and then gradually decreasing. For deterministic plants, the value could eventually be zero after all state-action pairs are experienced at least once, because the state-action value can accurately learn rewards of individual state-action pairs by only one time searching. But for stochastic or time-varying plants, the value of ϵ should always keep a small but non-zero value, in order to have more accurate estimations regarding external disturbances and track time-varying dynamics.

Besides the ϵ-greedy algorithm, there are also several other algorithms that aim to balance the trade-off between exploration and exploitation, such as the ϵ-soft algorithm [Tho97] and softmax algorithm [Bar98] [IYY02].

- Q-learning:

  The update rule for the state-action value function in Q-learning is defined as [Bar98]

$$\begin{aligned} Q(S(k),\mathbf{U}(k)) =& Q(S(k),\mathbf{U}(k)) + \alpha(k)\Big[R(k) \\ &+\gamma \max_{\mathbf{u}'} Q(S(k+1),\mathbf{u}') - Q(S(k),\mathbf{U}(k))\Big]. \end{aligned} \tag{4.48}$$

The only difference between Q-learning and sarsa is the definition of the TD error, which in Q-learning is [Bar98]

$$\delta_{\text{td}} = R(k) + \gamma \max_{\mathbf{u}'} Q(S(k+1), \mathbf{u}') - Q(S(k), \mathbf{U}(k)). \quad (4.49)$$

This difference reflects two different update principles of TD methods, which are called on-policy and off-policy. Sarsa is an on-policy algorithm because the update rule (equation 4.44) strictly follows the control policy π and all data used in equation 4.44 are actually experienced by the plant. Q-learning is an off-policy algorithm because the term $\max_{\mathbf{u}'} Q(S(k+1), \mathbf{u}')$ used in the update rule might not be the real experience of the plant. For example, if both sarsa and Q-learning are coupled with the greedy algorithm as the control policy, there is no difference. But if they use the ϵ-greedy algorithm as the control policy, the performance will be different. Because in Q-learning the update still follows a greedy algorithm, while in sarsa the update strictly follows the ϵ-greedy algorithm and the influence of exploration actions will be reflected in the update result. Compared between Q-learning and sarsa, Q-learning is in general more efficient because of the use of full knowledge rather than following the control policy.

- Actor-critic (AC) methods:

  AC methods are significantly different with sarsa and Q-learning [PS08b]. In both sarsa and Q-learning, the control policy is obtained based on the state-action value function, using stochastic searching algorithms such as the greedy algorithm or ϵ-greedy algorithm. Evaluation of the value function is the key of the entire control system and the control policy is only updated accordingly. This kind of algorithms is usually called critic-only algorithms (critic is the structure where the value function is updated).

  In AC methods, there are two separately structures which update the control policy and the value function independently, called the controller (actor) and the critic respectively (such as figure 4.9). The value function is updated in the critic using the same method as in Q-learning or sarsa. Meanwhile, the control policy is also updated in the controller. Instead of the probability distribution directly inferred from the value function using equation 4.43, the control policy used in AC methods is a parameterized family. The parameters of the control policy are updated either directly using the TD error, such as the Gibbs softmax algorithm [BV03] [PL05] and the adaptive heuristic critic learning architecture (AHCON) [Lin93], or using the gradient term that is influenced by the critic, like the policy gradient method [SMSM99] and the natural policy gradient method [PVS05] [PS06] [PS08a].

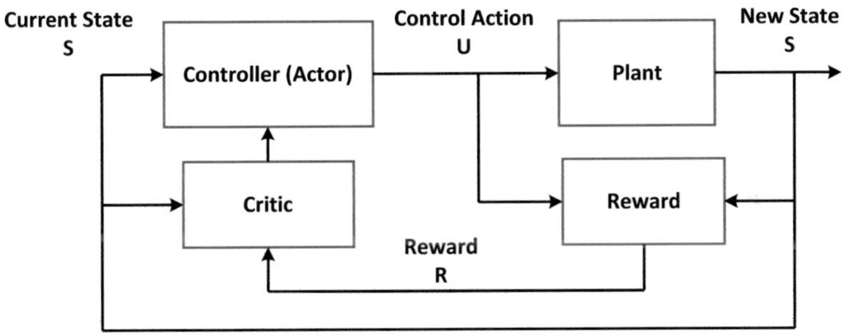

Figure 4.9. Actor-critic control system [GBLB12].

The structure of the AC control system seems similar with the conventional feedback controller (such as figure 4.2). To be more specific, the principle of AC methods is very close to the semi-direct SPSA based NN controller proposed previously (see figure 4.6). Despite minor differences with respect to different forms of realizations, both controllers are updated using terms either from direct interactions with the plant (TD error or direct SPSA) or gradients approximated from direct interactions (gradient methods or indirect SPSA). The critic use in AC methods can be represented by a lookup table such as in Q-learning or sarsa, but it can also be approximated using either linear or nonlinear (such as NN) functions [BM95] [Sut96] [SK00], which is similar with the system estimator in semi-direct SPSA NN control system. From another aspect, AC methods can also be considered as the combination of a critic-only algorithm and a model-free controller [LV09].

One advantage of AC methods over classical Q-learning and sarsa is that it is able to deal with continuous states and actions. Both the control policy and the value function in AC methods can be generalized in the continuous states and actions spaces by using various function approximation methods [SK00], respectively.

After its release in the 1980s, TD immediately became one of the most popular RLC methods. Until now many improvements have been done based on conventional TD methods (as introduced above). For instance, that function approximation based approaches have been proposed to extend the conventional critic-only methods into more general methods which can deal with continuous states and actions, such as the CMAC based Q-learning [SSR97] or the wire-fiited NN based Q-learning [GWZ99]. More detailed introductions about TD learning methods refer to [Boy02] and [Si04].

4.2.2. Design of Reinforcement Learning Controller

Although TD methods have been widely implemented in various applications, most of systems being controlled have simple architectures, which means they have either discrete states and actions, or low input and output space dimensions. When a TD learning controller is used to deal with complex systems with both continuous variables and high dimensions (such as HEPHAISTOS), the control system becomes much more complicated and in most cases the corresponding learning process will become extremely slow. It is acceptable if the controller can be trained with experimental data episodes in the offline form. However, in HEPHAISTOS it is impossible to generate training data which can comprehensively cover all dynamics of the plant. In this case, the TD learning control system has to be specially designed and optimized. Generally it is developed and implemented according to the following principles.

Hybrid multi-agent control structure

The hybrid multi-agent control structure is shown in figure 4.10. The entire control system consists of two independent controllers that are designed for different control objectives. One is a conventional adaptive controller that could use either MPC or NNC, and the other one is a lookup table based TD learning controller which will be introduced

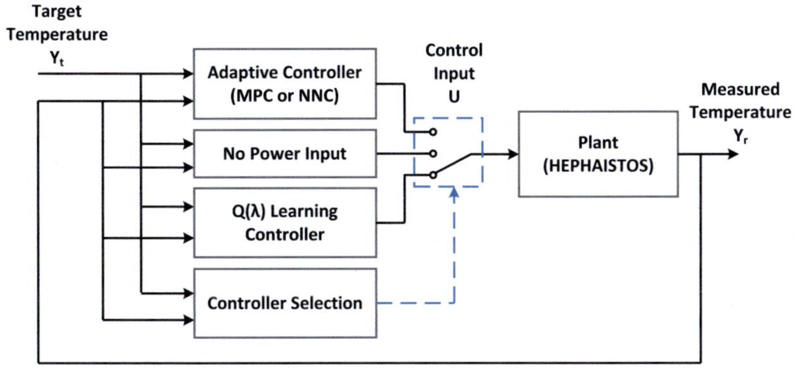

Figure 4.10. Hybrid TD learning control system.

later. This special hybrid control structure is developed because of several reasons.

First, in most control tasks the target temperature is varying along time such as shown in figure 4.11. Apparently, a randomly initialed TD learning controller has the worse ability to follow the target temperature change than conventional adaptive controllers. Because it does not use the same system estimation approach as in conventional adaptive control systems and it has to take a large number of exploration actions to get the accurate state values. Besides, it is not recommended to use a varying target in TD learning controllers, and the highly varying and stochastic environment/plant will also influence its learning results. Based on these reasons, it is better to use a conventional adaptive controller to handle the raising-temperature period (the blue part in figure 4.11).

Second, according to numerous experimental data (can be found in the next chapter), no matter which control method is applied, the controlled temperature distributions of different control algorithms are very similar to each other during the first raising-temperature period. All control algorithms have to control based on inaccurate system estimations and no effective control actions can be made. In other words, as long as the target temperature curve is defined, it is not likely to improve the temperature distribution on the first raising-temperature period. Compared with the first raising-temperature period, what is more important to the final temperature distribution is the flat-temperature period (the second red part in figure 4.11). In this case, the TD learning controller is perfectly suitable because of its ability to deal with intelligent control tasks and highly nonlinear plants.

Finally, the entire control task of the TD learning controller is significantly simplified by the hybrid control structure, from following a varying temperature in a large temperature range to converging to a fixed temperature in a small temperature range. As a result, the structure of the TD learning controller is also simplified and correspondingly the whole learning process is shortened.

Lookup table based $Q(\lambda)$-learning

After defining its applicable region, the next step is to determine the form of the TD learning controller. In principle, all kinds of TD learning controllers could be applied in HEPHAISTOS. An intuitive way to generate the MDP is to treat the temperature value as the state variable and the control input as the action variable. Both state and action variables are all continuous and it follows logically that the function approximation based AC methods (both the critic and the actor use the function approximation approach) should be implemented. However, it is also natural to convert the continuous states and actions into dis-

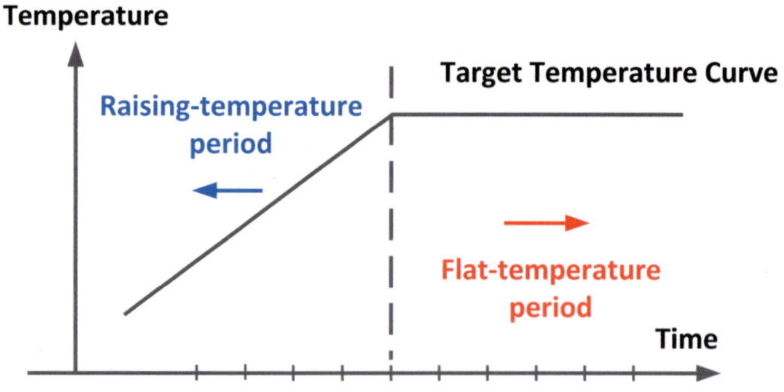

Figure 4.11. Target temperature curve during the heating process.

crete using either discretization [DKS+95] or fuzzy descriptions [LJ00] [Lin03], and then apply the lookup table based TD learning such as Q-learning and sarsa. After some preliminary tests and comparisons, the lookup table based Watkins' $Q(\lambda)$-learning [WD92] is selected as the TD learning control method.

The aforementioned sarsa (equation 4.44) and Q-learning (equation 4.48) are the most basic one-step TD methods that only update the value function of the current state or state-action pair based on the next state or state-action pair. At each time only one state or state-action value is updated and the overall convergence speed is limited. In order to fully exploit the usefulness of each reward and speed up the entire learning process, a more efficient TD(λ) learning method is developed [Tes95] [WS98]. The parameter λ refers to the use of an eligibility trace [LS98], which is normally defined as

$$e(s,\mathbf{u}) = \begin{cases} \gamma\lambda e(s,\mathbf{u}) + 1, & \text{if } s,\mathbf{u} \text{ is the current state-action pair,} \\ \gamma\lambda e(s,\mathbf{u}), & \text{Otherwise,} \end{cases} \tag{4.50}$$

where γ is the same as in equation 4.44 or 4.48 and $0 \leq \lambda \leq 1$ is the fading factor. Intuitively, the eligibility trace is considered as the temporary memory of the occurrence of each state-action pair. Each time when there is a reward, not only the current state-action pair but also former implemented state-action pairs should be assigned credit. The relevance between each state-action pair and the current reward is adjusted using the fading factor λ, indicating the fact that the relevance is decaying exponentially and the current state-action pair takes the main credit.

The involvement of the eligibility trace makes the TD(λ) method a combination of MC and pure TD (or TD(0)). When $\lambda = 0$, there is no former state-action pair recorded and then the method is equivalent to pure TD. When $\lambda = 1$, all former state-action pairs are recorded and the memory never fades, which means all state-action pairs take the same weight of credit from the current reward. Then the method becomes a online version of MC method. When the eligibility trace is combined with Q-learning, there are mainly two different approaches, the so-called Watkins' $Q(\lambda)$ and Peng's $Q(\lambda)$ [PW96]. Here the Watkins' $Q(\lambda)$ learning algorithm is selected because it is practically easier to be implemented. Its procedures are described in figure 4.12.

Initialize $Q(s, \mathbf{u})$ arbitrarily for all state-action pairs;
Initialize $e(s, \mathbf{u}) = 0$ for all state-action pairs;
Take the first control action $\mathbf{U}(k)$ at time $k = 1$ using the ϵ-greedy control policy;
At each time k $(k > 1)$:
repeat
Receive the reward $R(k-1)$ and observe the current state $S(k)$ based on the last state-action pair $(S(k-1), \mathbf{U}(k-1))$;
Update the eligibility trace matrix by $e(S(k-1), \mathbf{U}(k-1)) = e(S(k-1), \mathbf{U}(k-1)) + 1$;
Update the temporal difference by $\delta_{\text{td}} = R(k-1) + \gamma \max_{\mathbf{u}'} Q(S(k), \mathbf{u}') - Q(S(k-1), U(k-1))$;
Update the Q values for all state-action pairs by $Q(s, \mathbf{u}) = Q(s, \mathbf{u}) + \alpha \delta_{\text{td}} e(s, \mathbf{u})$;
Choose the control action $\mathbf{U}(k)$ from the current state $S(k)$ using the ϵ-greedy control policy;
If $\mathbf{U}(k)$ is the greedy control action, then $e(s, \mathbf{u}) = \gamma \lambda e(s, \mathbf{u})$ and otherwise $e(s, \mathbf{u}) = 0$;
until *the end of the control process;*

Figure 4.12. Procedures in the Watkins' $Q(\lambda)$ learning control.

The lookup table based $Q(\lambda)$ is selected as the TD learning control method here mainly under the consideration of control stability. The lookup table based TD learning is generally more stable than the function approximation based AC methods. On the one hand, TD learning with linear feature based approximation functions has been proved to converge to the optimal control policy [TVR97], but it is difficult to approximate the dynamics of HEPHAISTOS using linear features, especially to cover the part that how different control actions influence the state-action values . On the other hand, TD methods with nonlinear function approximations can easily become unstable during the learning process [B+95] [PSD01]. In contrast, the lookup table based

TD learning always converges to the optimal control policy with the appropriate learning rate (equation 4.46) and additional assumptions [Bar98]. Therefore it is more stable to be applied in the practice.

Empirical knowledge based discretization and optimization

Most lookup table based TD learning methods suffer the curse of dimensionality [Sut96], which means the overall computational time and memory of the learning grows exponentially with the the dimension of states and actions space. For example, it is assumed that the temperature vector consists of temperatures of 5 different locations and the control input vector contains inputs of 6 different heating sources. Each temperature value takes 3 possible states (temperature high/middle/low) and each control input variable takes 2 possible states (switch ON/OFF). The number of overall possible state-action pairs of the MDP is $3^5 \times 2^6 = 15552$, which means the lookup table has to estimate and save 15552 different Q values. If the control period is 1 s, then it needs at least 15552 s to go through all state-action pairs, not even mentioning to get the accurate estimation of each individual state-action values.

As stated in [BM03], empirical knowledge based options and policies are great approaches to accelerate the learning and provide guarantees about the system performance during the learning. In order to make the TD learning controller more realistic in practice, empirical knowledge based optimization is employed to further simplify the entire MDP and the corresponding $Q(\lambda)$ learning controller. Since the complexity of the controller mainly depends on the state and the action spaces, the simplification focuses on the reduction of dimensions of both the state and the action spaces.

Recalling the control task represented in equation 4.30 which aims to control N measured temperatures converge to the target temperature and meanwhile reduce the temperature window between $Y_{\max}$ and $Y_{\min}$. To reduce the number of the state space, this control task can be modified into a simpler task such as control $Y_{\max}$ and $Y_{\min}$ converge the target temperature. Because the modified control task is a necessary condition for the original control task. In other words, the overall temperature distribution will be improved if both $Y_{\max}$ and $Y_{\min}$

are converging to the target temperature. Using this control task, the number of controlled temperatures can be reduced to 2. For each of them, the number of possible states can also be reduced. For example, in many practical applications $Y_{\max}$ and $Y_{\min}$ does not necessarily to reach the target temperature, and a more realistic objective is to constrain them within a range as $[Y_{\text{tar}} - 5, Y_{\text{tar}} + 5]$. In this case, both $Y_{\max}$ and $Y_{\min}$ can be assigned with only 2 possible states, such as

$$\text{State of } Y_{\max} \qquad S_{\max} = \begin{cases} 1, & \text{when } Y_{\max} <= Y_{\text{tar}} + 5, \\ 0, & \text{Otherwise,} \end{cases} \tag{4.51}$$

and

$$\text{State of } Y_{\min} \qquad S_{\min} = \begin{cases} 1, & \text{when } Y_{\max} >= Y_{\text{tar}} - 5, \\ 0, & \text{Otherwise.} \end{cases} \tag{4.52}$$

This discretization method abstract all temperature distribution into 4 (2×2) different states, which significantly simplifies the whole state space. Due to the limited discretization resolution, several situations are not able to be well represented by the state. Therefore additional controller selection mechanism has to be added such as

$$\text{The active controller} \Leftarrow \begin{cases} \text{MPC/NNC}, & \text{when } Y_{\max} <= Y_{\text{tar}}, \\ \text{No Power}, & \text{when } Y_{\min} >= Y_{\text{tar}}, \\ Q(\lambda) \text{ learning}, & \text{otherwise} \end{cases} \tag{4.53}$$

However, unlike other measured temperatures, the positions of the maximum and the minimum temperatures are not fixed and varying along the heating process. In order to successfully control both of them, the information about their positions should also be discretized and included in the state variable. For instance, a rectangular workpiece is heated in the HEPHAISTOS cavity and its thermal picture can be shown as in figure 4.13. The whole heating area is divided into 5 different regions (R1, R2, R3, R4, R5) and each region represents one possible state for the position of $Y_{\max}$ and $Y_{\min}$. In addition, according to the empirical knowledge, the minimum temperature $Y_{\min}$ never occurs in the center region of the workpiece (R3), therefore the number of possible states for $Y_{\min}$ is 4. The number of total position states is $5 \times 4 = 20$.

Based on the discretization schemes of temperatures and positions, the overall number of possible states for the modeled MDP should be $4 \times 20 = 80$. This number could be further reduced using the state abstraction strategy [Die99]. The control objective is to constraint both $Y_{\max}$ and $Y_{\min}$ in the predefined temperature range. As long as they are in the range, the states of positions of $Y_{\max}$ and $Y_{\min}$ become less important. Therefore all states with $S_{\max} = 1$ and $S_{\min} = 1$ can be combined into one state. Although this state abstraction makes state-action value related with this state more stochastic, the number of total state-action pairs can be largely reduced due to this abstraction. As a result, the number of total states after the state abstraction is $|S| = 20 \times 3 + 1 = 61$.

Besides the states, the number of total actions is also reduced based on the empirical knowledge. In the new HEPHAISTOS cavity 3, there are 18 microwave heating sources. Apparently there is no way to include all of them in the $Q(\lambda)$ learning controller (2^{18} actions). Considering the applicable scenario of this controller is the flat-temperature period where no high power heating is needed, several rules are made to screen for the suitable control actions as following.

1. Each HEPHAISTOS module has 6 heating sources, but only 4 of them are used. Heating sources no. 5 and no. 8 (see figure 2.5) are

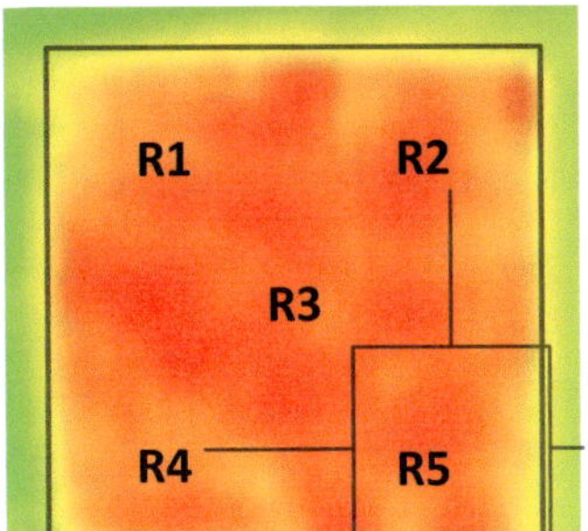

Figure 4.13. Area discretization.

neglected. Because they locate below the setup and their heating effects are weaker than the other 4. In total 12 different heating sources are used.

2. All control combinations consisting of more than 2 activated sources are neglected, because 2 heating sources (maximum 4 kW power) is enough to keep the workpiece of middle sizes ($1\text{m}^2 \sim 2\text{m}^2$) at the target temperature range ($70°\text{C} \sim 100°\text{C}$).

3. After the first two screenings, all left control combinations are further tested and the combinations that only heat the center region of the workpiece are eliminated.

Following the above three rules, 30 different control actions are finally selected to generate the action space. Combining the state and the action spaces together, there are entirely 1830 (61×30) different Q values to be estimated and saved by the $Q(\lambda)$ learning controller. In theory, even the control period is 1 s, it need at least 1830 s to go through the overall state-action space. But in practice, the hot spots and cold spots are not moving through the whole workpiece during one heating process. It often happens that the states of the system stay in a limited number of states and never go to other states, which means the number of practically possible state-action space is much smaller than the theoretical value. In this case, the control task is realistic to be solved in real time (see the results shown in chapter 4). The complete control procedures of the hybrid multi-agent control system are shown as in figure 4.14.

Define the state space S and the action space U;
Initialize $Q(s, \mathbf{u})$ arbitrarily for all $s \in S$, $\mathbf{u} \in U$;
Initialize $e(s, \mathbf{u}) = 0$ for all $s \in S$, $\mathbf{u} \in U$;
Initialize the adaptive controller $C_{\text{ad}}(\mathbf{Y}, Y_{\text{tar}})$ (either MPC or NNC);
Take the first control action $\mathbf{U}(k)$ at time $k = 1$ based on the equation $\mathbf{U}(k) = C_{\text{ad}}(Y, Y_{\text{tar}})$
At each time k $(k > 1)$:

repeat

  Select the active controller $C_{\text{act}}(k)$ according to

$$C_{\text{act}}(k) \Leftarrow \begin{cases} C_{\text{ad}}(\mathbf{Y}, Y_{\text{tar}}), & \text{when } Y_{\max} <= Y_{\text{tar}}, \\ \text{No Power}, & \text{when } Y_{\min} >= Y_{\text{tar}}, \\ Q(s, \mathbf{u}), & \text{otherwise.} \end{cases}$$

  if $C_{\text{act}}(k) = Q(s, \mathbf{u})$ **then**

    Calculate the control action $\mathbf{U}(k)$ and update the Q values according to the $Q(\lambda)$ learning algorithm (see figure 4.12);

  else

    Calculate the control action $\mathbf{U}(k) = C_{\text{act}}(k)$;

    Reset the eligibility trace matrix $e(s, \mathbf{u}) = 0$ for all $s \in S$, $\mathbf{u} \in U$;

until *the end of the control process;*

Figure 4.14. Procedures in the hybrid multi-agent $Q(\lambda)$ learning control system.

5. Experimental Results

In this chapter, the results of practical experiments will be presented. In the beginning, a number of preliminary trials are performed in order to verify the validity of the linear and the nonlinear models (given by the equation 3.41 and 3.46 respectively). Then the system identification algorithms introduced in chapter 3 are tested and compared, to identify the most appropriate system identification method for practical heating experiments. In the end, both the adaptive and the intelligent control methods are applied, and the corresponding experimental results are presented and compared.

5.1. Verification in HEPHAISTOS

Due to its complicated structure, some fundamental characteristics of HEPHAISTOS are difficult to determine theoretically, e.g. the properties of the EM field distributions. In order to understand the capabilities and limits of the HEPHAISTOS system better, a number of preliminary tests have been done with respect to one fundamental question, which is the practical power superposition principle.

According to the introductions in chapter 3, there are two different power superposition principles, defined as the scalar addition and the vector addition principles. These two principles are the basis of the linear 3.41 and the nonlinear 3.46 state-space models, respectively, which are also the foundations of the entire MPC system. From this point of view, to verify the power superposition principle is equivalent to the verification of validities of these two grey-box models.

The entire verification procedures contain three aspects. The first aspect is to check if each individual feeding source can provide a stable heating pattern. For each microwave source of HEPHAISTOS, whenever it is switched on, the frequency and phase of the output microwave is not fixed. If there is no stable heating pattern from each source, the whole microwave heating system will become a highly stochastic system where the future temperature is not predictable by neither of the two models.

If a stable heating pattern from each source can be guaranteed, the second part is to see if the combinations of any two or more feeding sources can provide stable heating patterns. This is important for the modeling. Because in practice when multiple feeding sources are switched on at the same time, unexpected coupling effects will occur between different sources. It is possible that when multiple sources are switched on simultaneously, the resulting superposed heating patterns differ a lot from time to time, which could also make the system identification process more difficult and the models unreliable. After the first two aspects are confirmed, the last part of the validation is to verify that if the scalar and the vector addition principles are able to describe the power superpositions correctly.

The setup used in the validations is shown as in figure 5.1. The temperature distribution of this setup was monitored by an infrared camera in real time. All following experiments were done at the same temperature range (30 °C ∼ 32 °C) and the temperature of surrounding air was also the same (19 °C ∼ 21 °C), therefore all temperature-dependent parameters can be considered as constants.

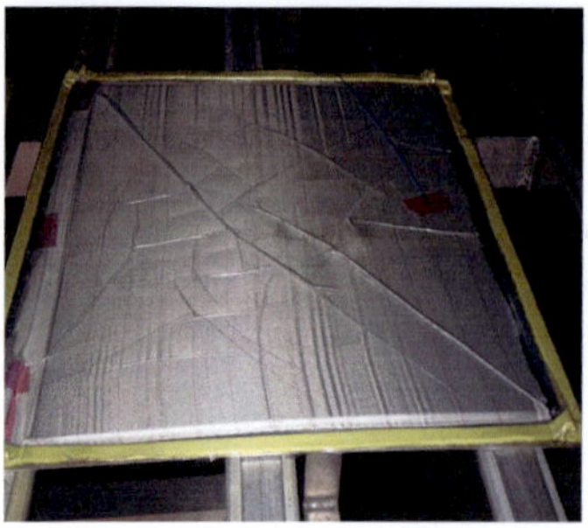

(a)

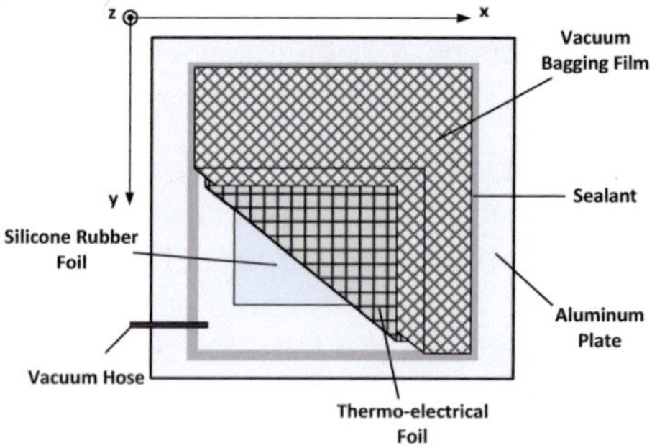

(b)

Figure 5.1. The setup used in verification experiments.

Step 1: Heating patterns of single feeding source

The heating stability verified in the experiment refers to two parts. The first part is to see how the heating pattern changes with time, and the other one is to test if the heating patterns are equivalent when the same source is switched on and off multiple times. If a feeding source has the same heating pattern at different trials and this pattern is stable over time, then it is guaranteed that this source can provide a stable power output as well as a stable EM field distribution.

For all different heating sources j in the first two modules (see figure 2.13, $1 \leq j \leq 12$):

repeat
    repeat
        Wait until the temperature of the workpiece is cooled down to $30\,^{\circ}\mathrm{C} \sim 32\,^{\circ}\mathrm{C}$;
        Switch the heating source j for 10 s;
        Record the heating pattern from the infrared camera at time $t = 0, 2, 4, 6, 8, 10$ s.
    until *for three times*;
until *all sources are tested*;

Figure 5.2. Procedures to test the heating patterns of single source.

The procedures for the experiment are described in figure 5.2. For individual sources, the typical results are illustrated as in figure 5.3 and 5.4. The numbers within figures 5.3 represent the temperature changes of this point at this time compared with the initial temperature in the first figure at $t = 0$ s. For example, the number $+0.60\,^{\circ}\mathrm{C}$ in location 3 at time $t = 2$ s represents that the temperature of this location increases $0.60\,^{\circ}\mathrm{C}$ during the last 2 s. Although the temperature increasings are not entirely linear, the whole temperature distribution (thermal pattern) stays the same during the 10 s, which indicates a stable heating pattern from this source.

Compared with this heating stability of time, what is more important is the stability of the heating patterns at different trials. Because in

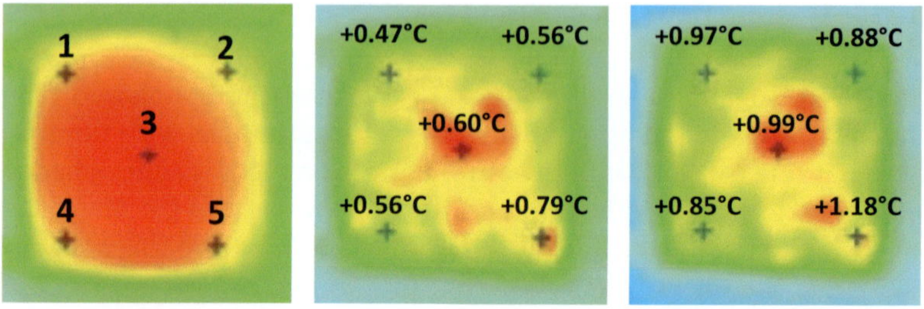

(a) $t = 0$ s (b) $t = 2$ s (c) $t = 4$ s

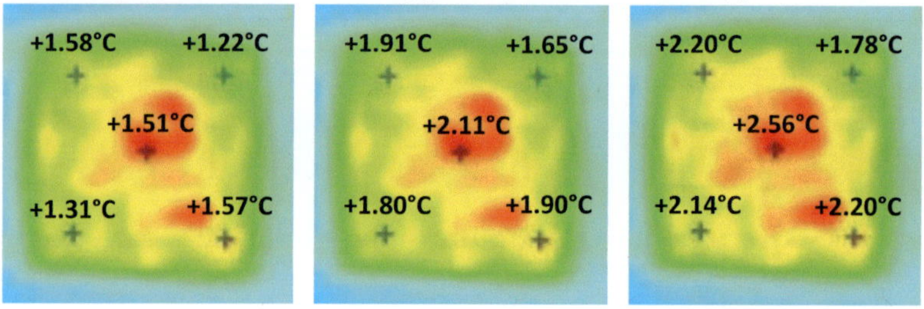

(d) $t = 6$ s (e) $t = 8$ s (f) $t = 10$ s

Figure 5.3. Heating patterns of source No. 3 (new CA3) at different time of the same trial.

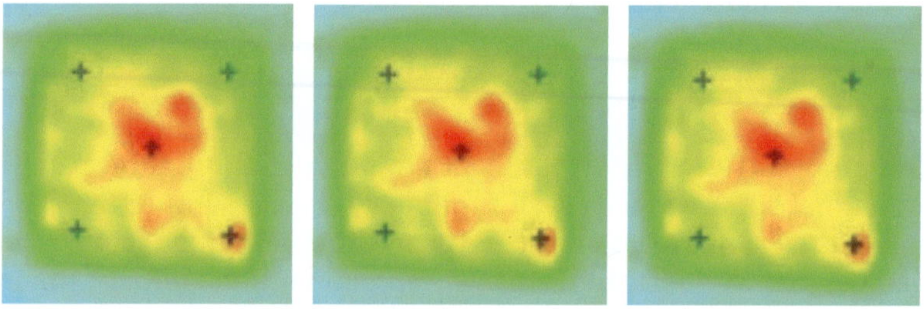

(a) Trial 1 at time $t = 2$s (b) Trial 2 at time $t = 2$s (c) Trial 3 at time $t = 2$s

Figure 5.4. Heating patterns of source No. 3 (new CA3) at the same time of different trials.

practical applications, the control input changes quickly and it seldom happens that a heating source is switched on for a long time. As shown in figures 5.4, the heating patterns of different trials are almost identical. The same results are also reflected by the heating rates as in figure 5.5.

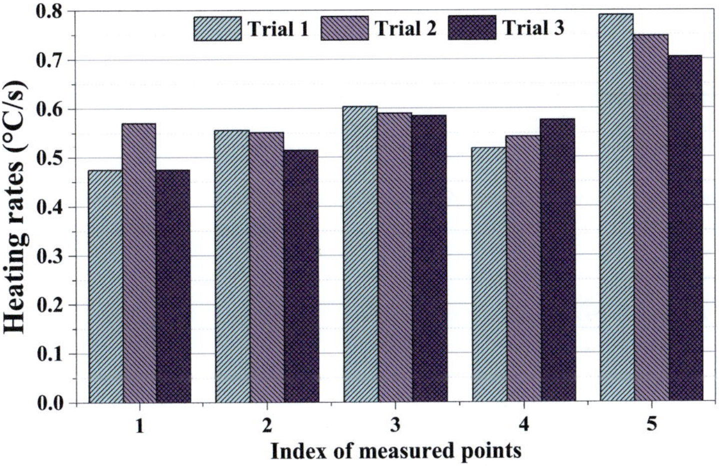

Figure 5.5. Heating rates comparison of 3 trials during the first two seconds.

The heating rates shown in figure 5.5 are calculated using the following equation

$$\mathbf{R} = \frac{1}{\Delta t}\big(\mathbf{Y}(k+1) - \mathbf{A}(k)\mathbf{Y}(k)\big) \;\propto\; \mathbf{P}_{\mathrm{mw}}. \tag{5.1}$$

Since the heating rates are directly proportional to the microwave heating power, therefore they can be used as representatives of the heating power. In other words, if the heating rates of different sources follow the scalar addition rule, it means the heating power also follows the scalar addition rule. This principle will be used later in the third step of the verification.

Step 2: Heating patterns of multiple feeding sources

The second step is to test the heating stability of multiple feeding sources. The procedures used in this setup is the same as described in figure 5.2, except the objective is switched to different combinations of heating sources. The verification results can be found in fi gures 5.6 and 5.7. Since the transient heating rates are more relevant to the practical applications, only results about the short time heating patterns (t = 2 s) are demonstrated and analyzed in the following.

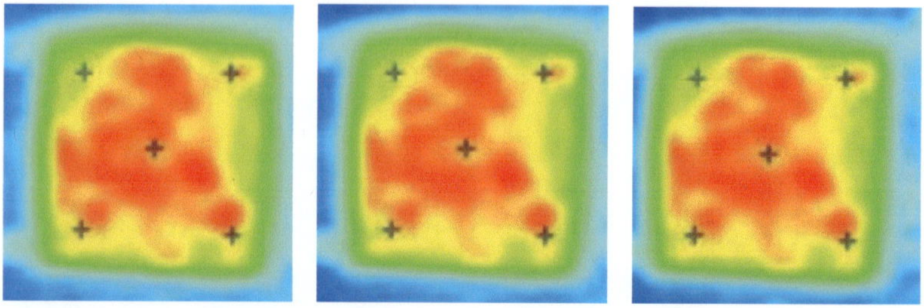

(a) Trial 1 at time t = 2s (b) Trial 2 at time t = 2s (c) Trial 3 at time t = 2s

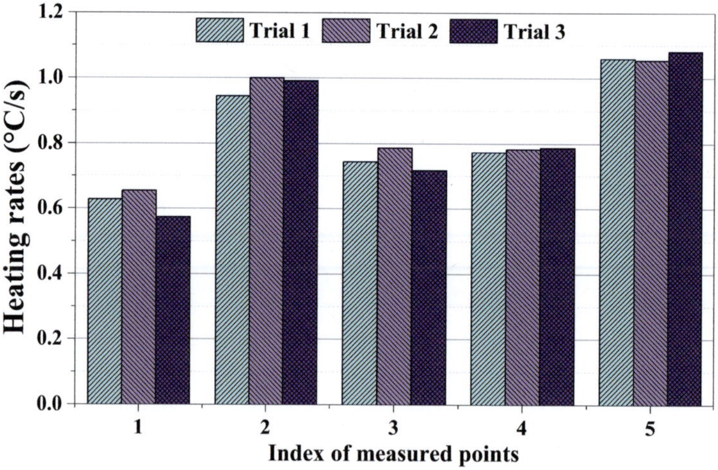

(d) Heating rates of 3 trials.

Figure 5.6. Heating patterns (a, b, c) and rates (d) of source No. 1 and 5 at 3 different trials.

The results in figures 5.6 and 5.7 both indicate that even multiple heating sources are switched on simultaneously, the corresponding superposed heating patterns are still stable. Meanwhile, the number of heating sources does not have any influences to this stability. Results from the first and the second steps prove that HEPHAISTOS is a non-stochastic system. The dynamics of HEPHAISTOS are possible to be modeled and learned via repeated experiments and tests, which provides a solid foundation to all previous introduced modeling and control methods.

Step 3: Validity of linear and nonlinear models

In the last verification step, the previously mentioned principle of heating rates is utilized, to verify if models 3.41 and 3.46 are accurate approximations of practical heating situations. In order to get a more comprehensive result, more temperatures were included in the following verification experiment such as in figure 5.8 . Corresponding Verification procedures are described as in figure 5.9.

It is easy to test the linear superposition rule, by comparing the linearly superposed heating rate $\mathbf{R}_{\text{linear}}$ with the real overall heating rate $\mathbf{R}_{\text{real}}$. If the linear superposed heating rate $\mathbf{R}_{\text{linear}}$ is equivalent to the real heating rate $\mathbf{R}_{\text{real}}$, it means that the linear superposed power also equals to the practical overall heating power, and consequently, it proves the validity of the linear model 3.41 . The corresponding verification result for two different feeding sources is shown in the following figure 5.10 .

From the above figure it is noted that the linear superposed heating rate $\mathbf{R}_{\text{linear}}$ is close to the real heating rate $\mathbf{R}_{\text{real}}$. For situations with more sources, the same result still holds 5.11.

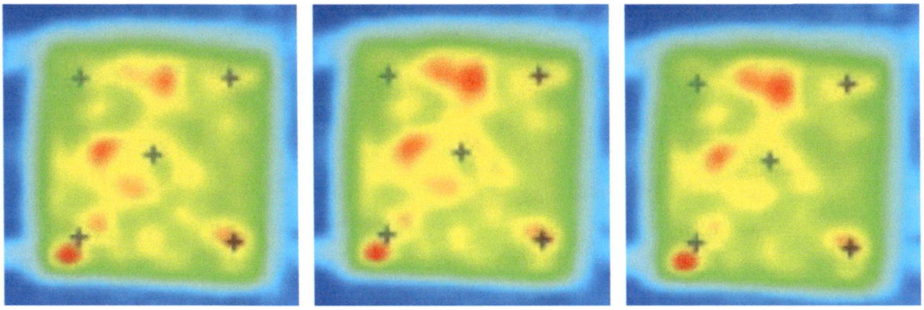

(a) Trial 1 at time t = 2s (b) Trial 2 at time t = 2s (c) Trial 3 at time t = 2s

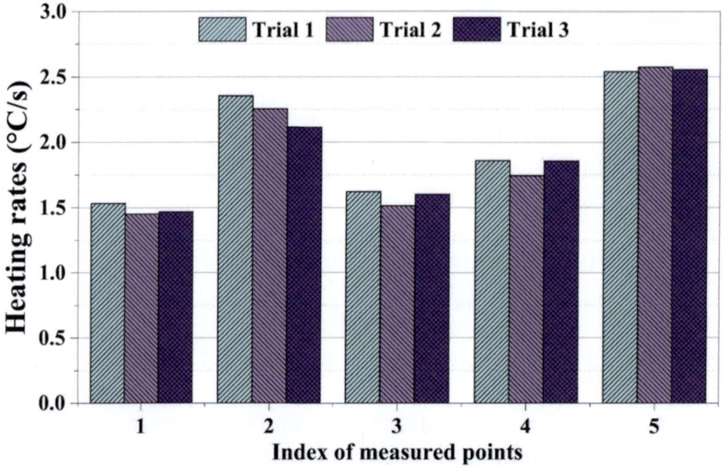

(d) Heating rates of 3 trials.

Figure 5.7. Heating patterns (a, b, c) and rates (d) of 12 sources (from 1 to 12) at 3 different trials.

Results in 5.10 and 5.11 are just epitomes of a large scale of experimental data. In most cases accuracies of the linear superposed heating rates are over 85%. The accuracies are even higher (over 90%) for for combinations of a large number of heating sources (more than 4). Therefore, in general the linear model 3.41 is considered as a accurate approximation of the practical heating scenarios. Nevertheless, on the other hand, in certain cases the differences between real and estimated heating rates can be drastic. such as the results shown in the following figure 5.12. It is clear that there are 4 hot spot areas in the heating pattern of No. 3 and 7 (as illustrated by the dashed circles in figure 5.12), three of which can be interpreted by the linear superposition principle. But the heating rate of the upper left corner is much higher than the linear superposed value (as index '1' in figure 5.12d). In fact, both the first two thermal patterns have low heating rates (cold spot) in the upper left corner, but the pattern of the heating combination has a clear hot spot in this region. Meanwhile, it is also worthy to notice that in many cases (including the case shown in figure 5.12), the thermal pattern of the heating combination is not exactly equiva-

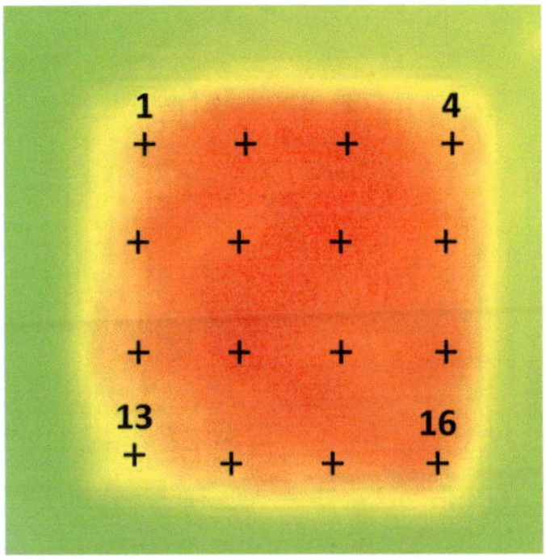

Figure 5.8. Measured points on the load.

Perform following trials at similar environments (the temperature of the load at about 30 °C, and the temperature of its surrounding air at about 20 °C);

repeat

  Randomly select m different heating sources and switch them on simultaneously for 2 s;

  Calculate the overall heating rate using the equation 5.1, denoted by $\mathbf{R}_{\text{real}}$;

  for *for each selected source* j **do**

    Switch each heating source j for 2 s;

    Calculate the individual heating rate using the equation 5.1, denoted by $\mathbf{R}_j$;

  Calculate the linearly superposed heating rate $\mathbf{R}_{\text{linear}}$, such as

$$\mathbf{R}_{\text{linear}} = \sum_{j=1}^{m} \mathbf{R}_j .$$

  Compare the real heating rate $\mathbf{R}_{\text{real}}$ and the linear superposed $\mathbf{R}_{\text{linear}}$.

until *the end of the process*;

Figure 5.9. Procedures of the verification of heating rate superpositions.

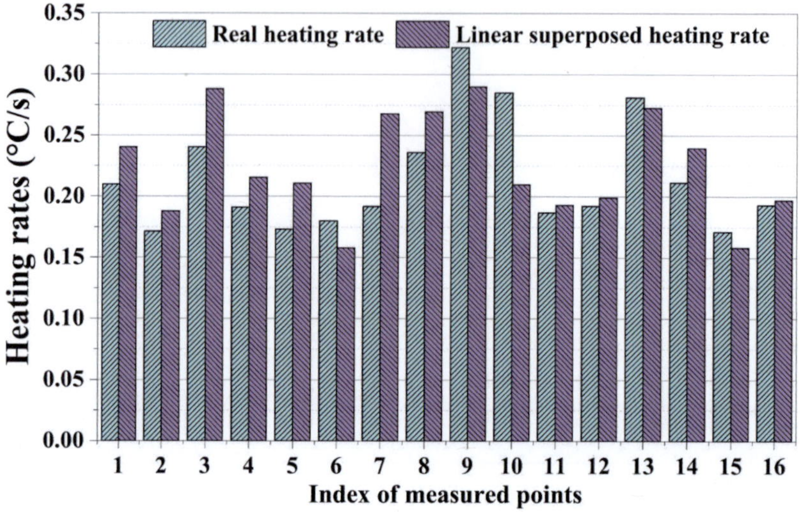

Figure 5.10. Comparison between real and linear superposed heating rates of 2 sources (No. 1 and 2) at 16 different points.

lent to the linear superposition of individual thermal patterns. For the case shown in figure 5.12, a legitimate explanation is that the EM field created by the sources No. 3 and 7 have constructive superpositions in the upper left corner and the corresponding heating power follows the vector addition rule. From this point of view, although the nonlinear system model 3.46 can not be directly verified in the same way as the linear model, it is still reasonable to use the nonlinear model and the vector addition principle to explain the thermal pattern superposition phenomena, especially for scenarios with a small number of different feeding sources.

According to all experimental results, a brief conclusion can be made as that the general heating power (rates) superposition within HEPHAISTOS can be regarded as a combination of both scalar and vector additions. The practical heating scenario is a varying combination of both the equations 3.41 and 3.46. It tends towards 3.41 when the number of feeding sources is large, because of stronger cross influences between multiple microwave generators, and towards 3.46

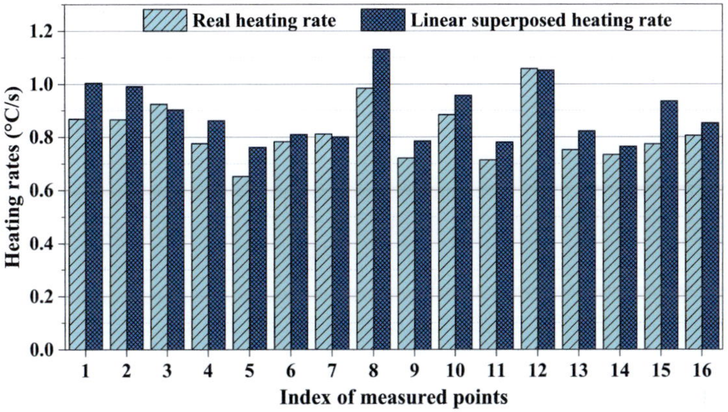

(a) Result of 4 feeding sources (No. 1−4).

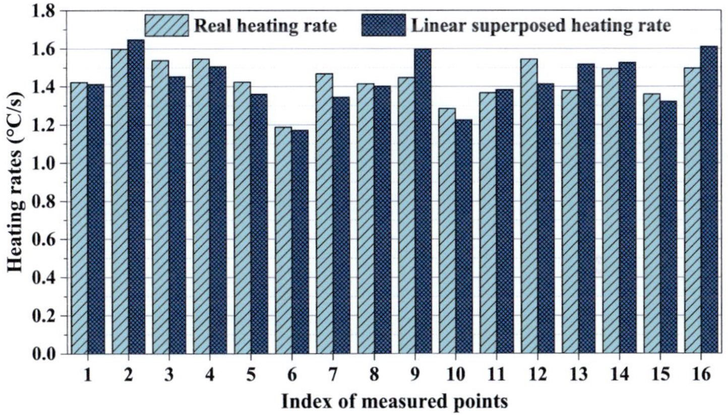

(b) Results of 8 feeding sources (No. 1−8).

Figure 5.11. Comparison between real and linear superposed heating rates of 4 (a) or 8 (b) feeding sources at 16 different points.

when the number of feeding sources is low, particularly for cases where sources are from different modules.

After above 3 steps, assumptions used in the grey-box modeling have been verified and guaranteed to be accurate to a great extent. In addition, heating patterns and corresponding heating rates obtained from the verification experiments are quite helpful in the following controller design, especially for the intelligent controller construction.

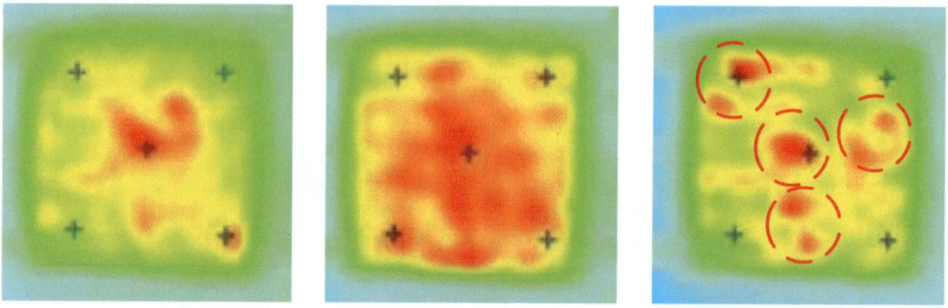

(a) Heating pattern of No. 3. (b) Heating pattern of No. 7. (c) Heating pattern of No. 3 and 7.

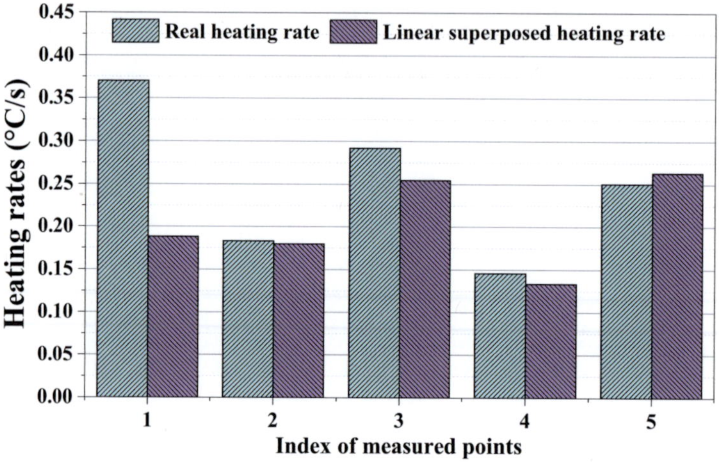

(d) Comparison of heating rates at 5 different points.

Figure 5.12. Comparison of Heating patterns (a, b, c) and rates (d) of No. 3 and 7.

5.2. Results of System Identification

Based on the verification results, different system identification algorithms have been tested and compared in this section, to identify which one is more accurate to estimate the real dynamics of the HEPHAISOTS system. Data from practical experiments are used to certify the reliability of each algorithm. The validation of system identification algorithms contain two different tests. The first test is to use different algorithms to learn the dynamics of the plant from the experimental data, and check their one-step prediction accuracies. The one-step prediction accuracy indicates the ability to predict the future temperatures based on the current temperatures and control input values, which is how the system estimator functions in real control process. The second test is the regeneration of temperature curves. Given the same initial temperature value and the whole control input sequence, the estimators trained by the first test are used to regenerate the entire temperature sequence. The regenerated data are compared with the real data, to test if individual algorithms really capture the dynamics of HEPHAISTOS.

It is assumed that the data used in two tests are given as

$$D_1 = \left\{ \left(\mathbf{U}_{\mathrm{d1}}(1), \mathbf{Y}_{\mathrm{d1}}(1)\right), \left(\mathbf{U}_{\mathrm{d1}}(2), \mathbf{Y}_{\mathrm{d1}}(2)\right), \ldots, \left(\mathbf{U}_{\mathrm{d1}}(Q_1), \mathbf{Y}_{\mathrm{d1}}(Q_1)\right) \right\},$$
$$D_2 = \left\{ \left(\mathbf{U}_{\mathrm{d2}}(1), \mathbf{Y}_{\mathrm{d2}}(1)\right), \left(\mathbf{U}_{\mathrm{d2}}(2), \mathbf{Y}_{\mathrm{d2}}(2)\right), \ldots, \left(\mathbf{U}_{\mathrm{d2}}(Q_2), \mathbf{Y}_{\mathrm{d2}}(Q_2)\right) \right\},$$

where D_1 and D_2 are data sets obtained from experiments using the same setup and the same heating environment. In the first test, the objective is to use each algorithm for online system identification (the data pair is presented and estimated one by one), and compare the one-step predicted temperature $\mathrm{Y}_{\mathrm{pre}}$ and the real temperature Y_{d1}. The one-step prediction is expressed as

$$\begin{aligned} \mathrm{Y}_{\mathrm{pre}}(1) &= \mathrm{Y}_{\mathrm{d1}}(1), \\ \mathrm{Y}_{\mathrm{pre}}(q+1) &= f_q(\mathrm{Y}_{\mathrm{d1}}(q), \mathbf{U}_{\mathrm{d1}}(q)), \quad 1 \leq q \leq Q_1 - 1, \end{aligned}$$

where f_q represents the function of the system estimator (after the update of the q-th data pair).

After the first test, the second test aims to regenerate the whole data set T_2 by

$$\mathrm{Y}_{\mathrm{reg}}(1) = \mathrm{Y}_{\mathrm{d2}}(1),$$
$$\mathrm{Y}_{\mathrm{reg}}(q+1) = f_{Q_1-1}(\mathrm{Y}_{\mathrm{reg}}(q), \mathbf{U}_{\mathrm{d2}}(q)), \quad 1 \leq q \leq Q_2 - 1,$$

The regenerated temperature $\mathrm{Y}_{\mathrm{pre}}$ will also be compared with the real data Y_{d2} to determine the regeneration error. Compared with the one-step prediction, the test of regeneration can more comprehensively reflect the estimation accuracy of each system identification algorithm, and hence, is an important factor to be taken into account for the determination of the final implemented system identification algorithm.

5.2.1. Grey-box Approaches

For the grey-box modeling approach, four different system identification schemes have been tested, including the nonlinear MISO EKF, the linear MISO RKF, the linear MISO RLS and the linear MIMO RKF. Differences among these algorithms can be found in chapter 3. The data used in these tests are obtained from experiments with eight heating sources and two controlled temperatures. Results of the different algorithms are shown in figures 5.13, and 5.14.

In order to get more reliable evaluations to the performance of each algorithm, each algorithm was tested for 20 times and the averaged performance is shown in table 5.1. In following tests, the sampling (discretization) period is $\Delta t = 1$ s, which means the processing time is equivalent to the number of data pairs (1 data pair per second).

The results shown in the figures can generally reflect the performance of individual algorithms in practical heating experiments. Several brief conclusions are obtained from above results.

- From the converging speed point of view, the nonlinear MISO EKF algorithm has the fastest converging speed among these four algorithms, using less than 100 seconds to reach the general MSE level of 0.1. In comparison, both MISO RKF and MIMO RKF took about 100 seconds to reach that MSE level. MISO RKF has the slowest converging speed, with about 450 seconds to reach that error level.

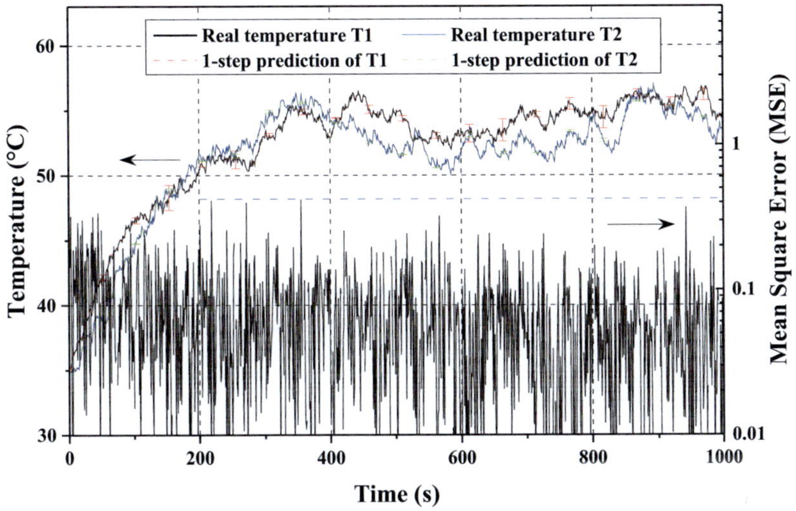

(a) One step predictions of nonlinear MISO EKF.

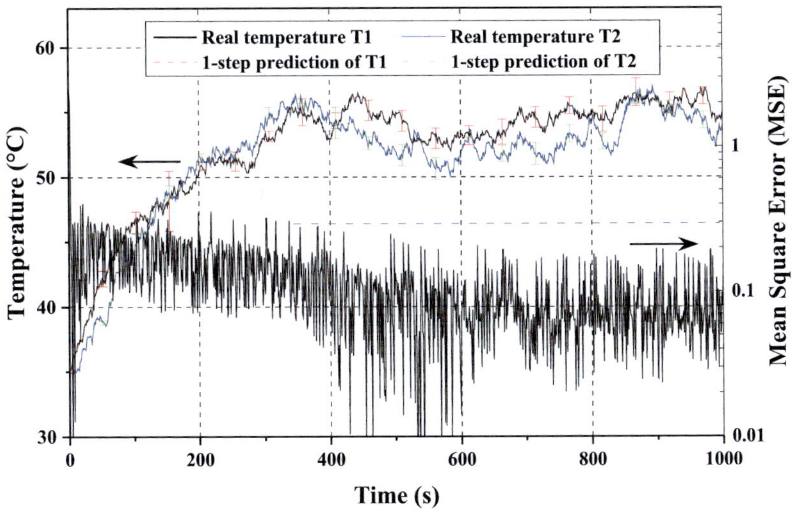

(b) One step predictions of linear MISO RKF.

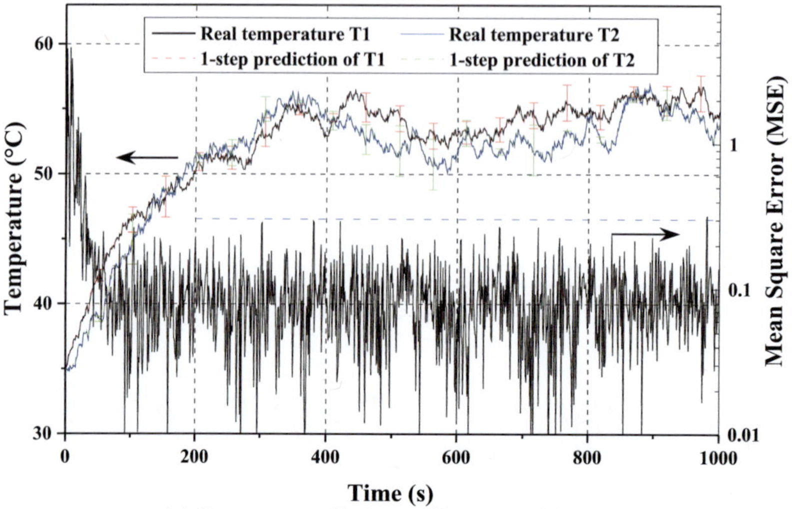

(c) One step predictions of linear MISO RLS.

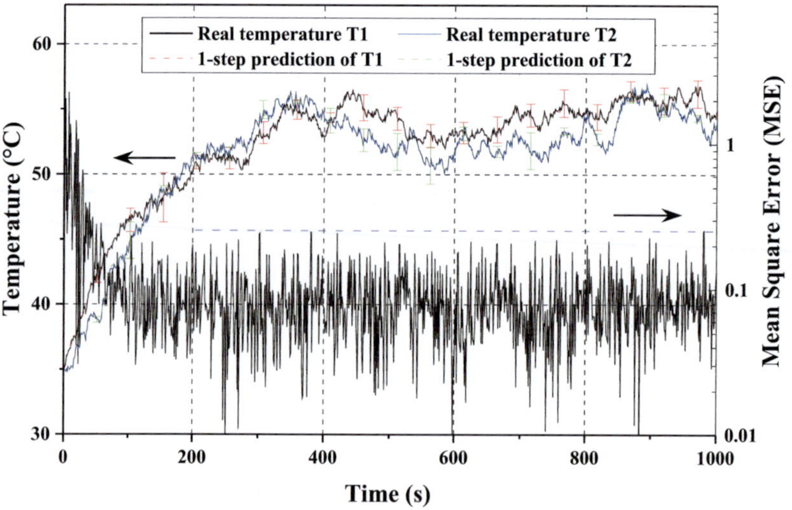

(d) One step predictions of linear MIMO RKF.

Figure 5.13. One-step prediction performance of four different system identification algorithms.

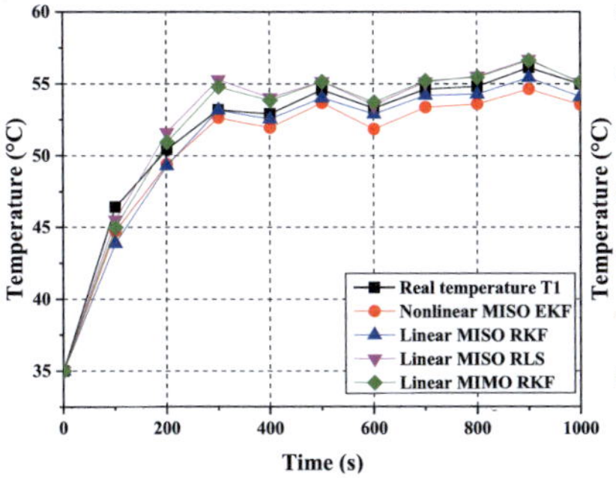

(a) Validation of T_1.

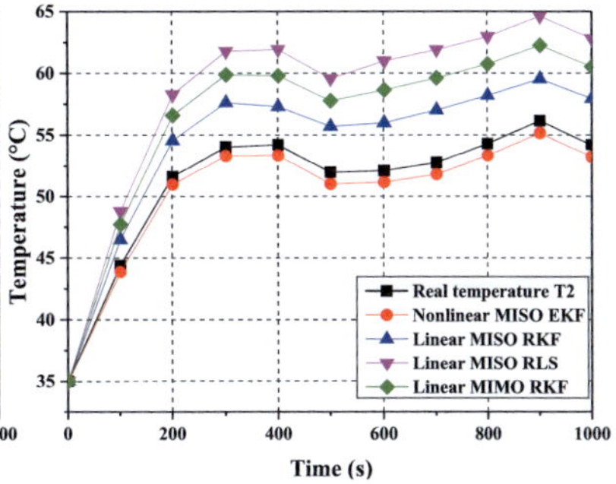

(b) Validation of T_2.

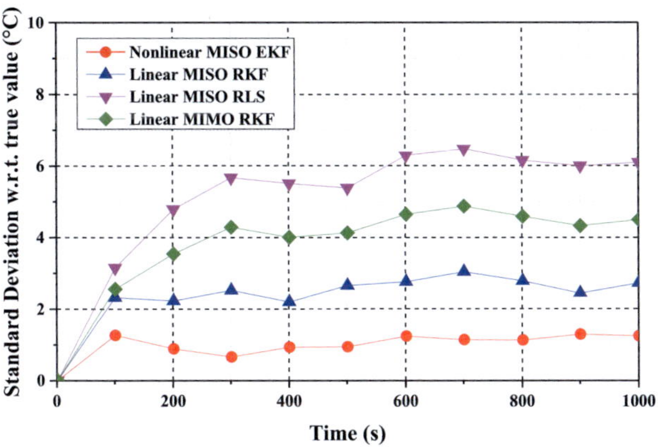

(c) Standard deviation (SD) with respect to the true values during the validation.

Figure 5.14. Validation of each algorithm using data of another experiment.

- In the accuracy aspect, the nonlinear MISO EKF algorithm has the minimum MSE for both the on-line system identification and the validation (see table 5.1). Compared between the other three linear algorithms, both the MISO RKF and MIMO RKF approaches are more accurate than the MISO RLS algorithm, which means the RKF algorithms is more suitable for HEPHAISTOS. On the other side, the MISO RKF is better than the MIMO RKF. The main reason is that in the MISO approach, only the diagonal elements of the matrix **A** have to be estimated, but in the MIMO approach the entire matrix is estimated. More estimation variables bring a larger estimation error and therefore degrade the estimation accuracy.

According to the above results, in practical experiments, the MISO RKF algorithm is used for the linear MPC and the nonlinear MISO EKF is used for the nonlinear MPC.

Algorithm	SID MSE	Mean SD of validation	Time (per call)
Nonlinear MISO EKF	0.0697	1.0698	0.00103 s
Linear MISO RKF	0.1205	2.5691	0.00150 s
Linear MISO RLS	0.1214	5.5467	0.00149 s
Linear MIMO RKF	0.1068	4.1431	0.00148 s

Table 5.1. Performance comparison of 4 grey-box SID algorithms.

5.2.2. Black-box Approaches

There are three algorithms introduced for the training of the NN, and their system identification performance regarding the real experimental data is shown in figures 5.15 and 5.16. As in the test of grey-box approaches, different algorithms were also tested for 20 times and the averaged parameters are shown in table 5.2.

Above results approximately reflect the performance of the three different learning in practice. SWBP has the minimum prediction MSE and a fast converging speed, but its validation error is the largest. More important, SWBP needs much more time compared with the other two algorithms. If the input or output dimensions increase, this execution time will become larger, which means it is not appropriate

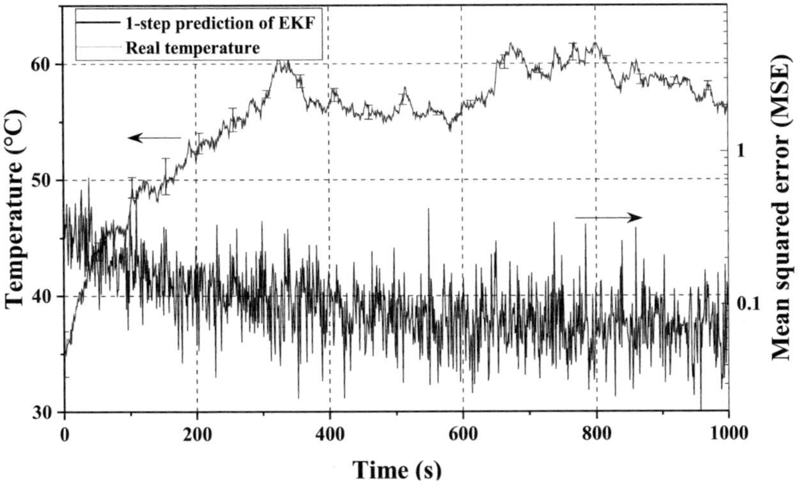

(a) One-step prediction results of EKF.

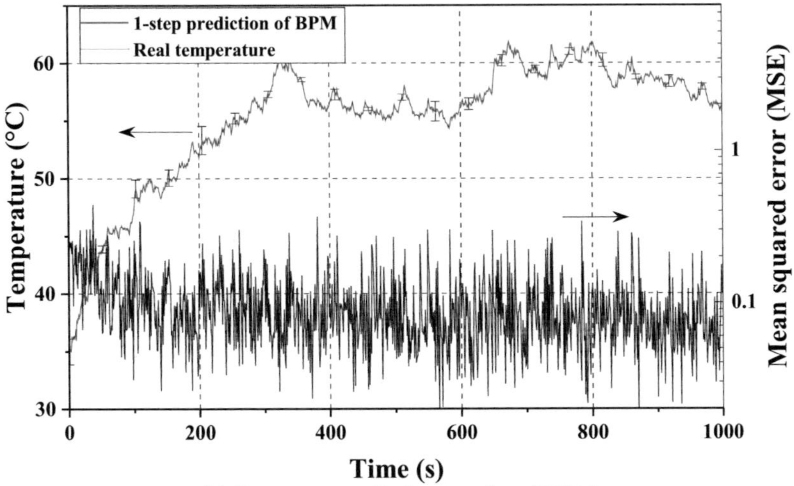

(b) One-step prediction results of BPM.

to be used in the practical fast and high-dimensional system identification tasks.

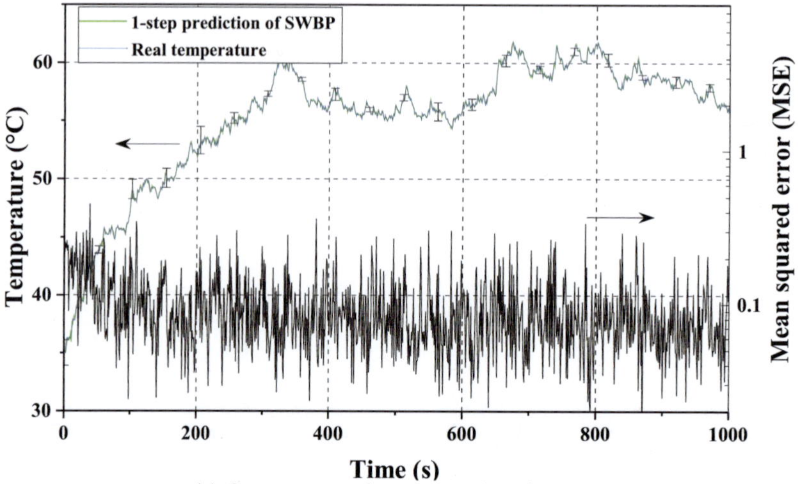

(c) One-step prediction results of SWBP.

Figure 5.15. One-step prediction results of three algorithms.

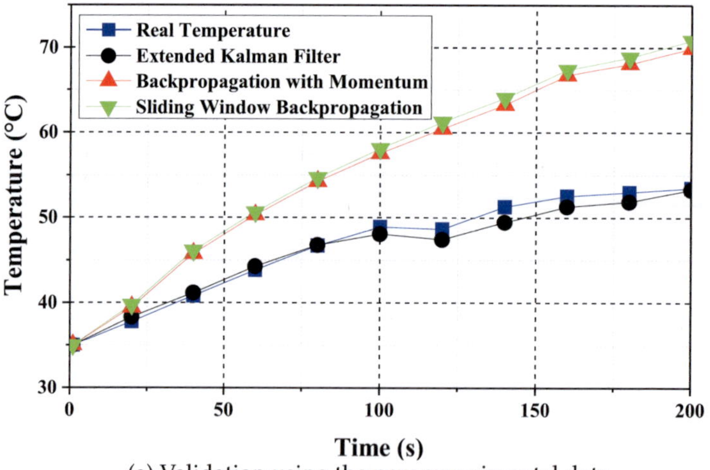

(a) Validation using the new experimental data.

Among these three algorithms, EKF has the minimum validation error, which means its true estimation accuracy is the highest. Meanwhile, its prediction accuracy and execution time are also acceptable. Due

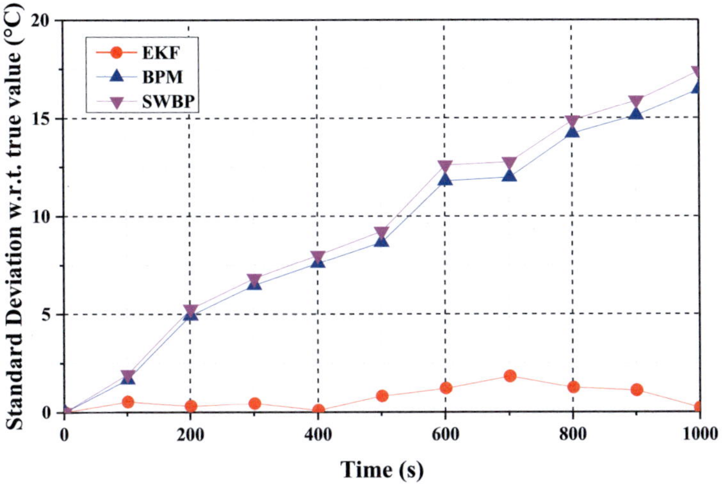

(b) Comparison of the standard deviation.

Figure 5.16. Validation results of three algorithms.

Algorithm	Learning MSE	Mean SD of validation	Time (per update)
EKF	0.1172	0.7743	0.0031 s
BPM	0.0990	9.8806	0.0015 s
SWBP	0.0974	10.4665	0.3446 s

Table 5.2. Performance comparison of three online NN training algorithms.

to the highest estimation accuracy and more comprehensive performance, the EKF algorithm is applied in the NNC system to train the system estimator.

Performance of both grey-box and black-box approaches can be compared qualitatively according to above results. On the one hand, both linear and nonlinear on-line recursive system identification algorithms have fast converging speed than the NN approaches. But on the other hand, the EKF based NN approach has a much smaller MSE than the nonlinear EKF algorithm in the validation part, which indicates the black-box approach is a more powerful tool to model systems with unknown dynamics.

5.3. Results of Different Control Methods

The performance of different control methods is the most important result of this dissertation. In this section, the control performance of all aforementioned control methods is presented, with respect to different setups. Before the experimental results of temperature control are demonstrated, an important question that has to be discussed is the controllability of the system.

Controllability is an important property of a system. According to the definition in [Dor95], a system is controllable if any initial state of the system can be moved to any other state in a finite time interval using the external input. Take HEPHAISTOS for example, it is controllable if any desired temperature distributions can be obtained within a finite time interval using the control input from the initial temperature distribution. Compared with this complete controllability, what is more important to the heating applications is the reachability of certain states in HEPHAISTOS. A particular state of the system is reachable if any initial states can be transfered to this state within a finite time intervals using a corresponding control input sequence [Rug96]. In HEPHAISTOS, the control task is to achieve a homogeneous temperature distribution for the whole workpiece, therefore the reachability of the homogeneous temperature distribution becomes the most important property of the system.

Although it has been proved in [WYT12] that any arbitrary temperature distribution could be achieved by a corresponding EM field distribution in microwave heating, it neglected that not all EM field distributions can be realized within a microwave cavity. Moreover, the EM field distribution of HEPHAISTOS is so far not possible to be obtained in neither analytical nor numerical ways. Due to these reasons, it is more realistic to analyze the controllability quantitively via a number of real experiments.

Through numerous practical heating experiments, the main conclusion can be obtained that the controllability of HEPHAISTOS is largely determined by the heating setup. To be more specific, different heating setups create different boundary conditions for different parts of the workpiece, which significantly affects the overall controllability. For example, for the setup shown in figure 5.1 with large sizes ($1.2\text{m} \times 1\text{m} \times 2\text{mm}$), it is inevitable to have hot spots in the central part of the workpiece and cold spots in the corners and edges. A typical heating process can be illustrated by the figures in 5.17.

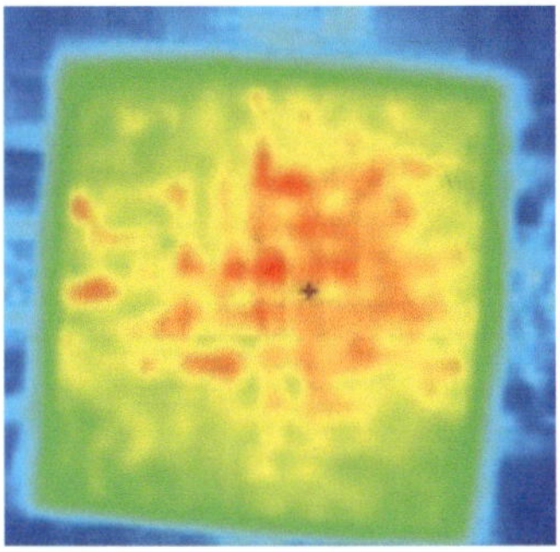

(a) t = 5 min.

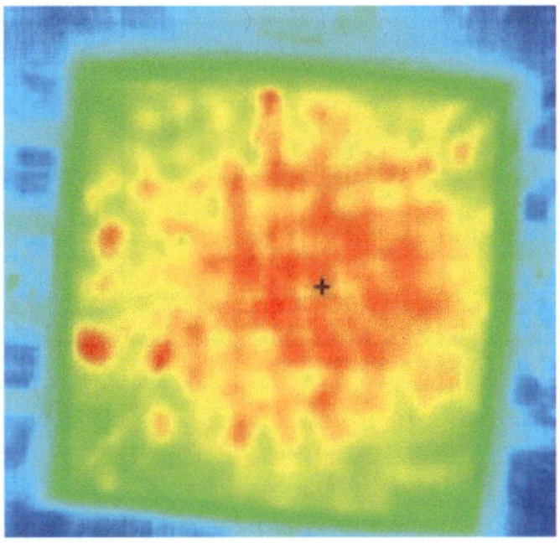

(b) t = 10 min.

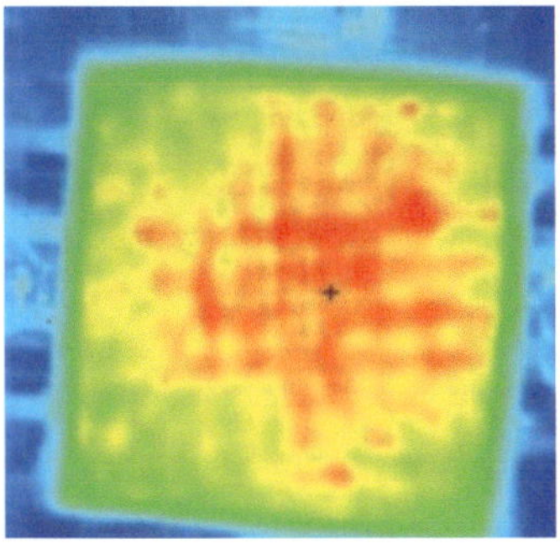

(c) t = 15 min.

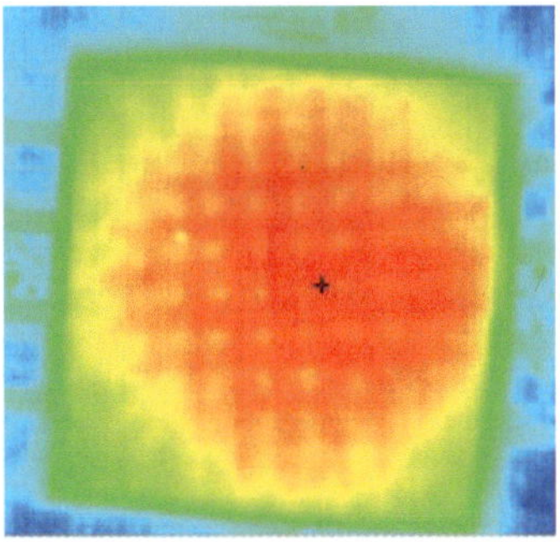

(d) t = 20 min.

Figure 5.17. Heating process (from (a) to (d)) with the setup of aluminum plate.

That is mainly due to different thermal conduction impacts from the aluminum plate and the metal table below the aluminum plate. During the heating process, the central part of the aluminum plate is always hotter than the corners and edges, because of less heat dissipation to the ambiance. Therefore for the workpiece, the heat dissipated to the aluminum plate through thermal conduction in the central part is always lower than the other parts, leading to a lower cooling rate and higher temperature. In addition, the heat convection effect of the central part is also lower than that of the other parts, because of the inhomogeneous temperature distribution of the surrounding air. As a result of above impacts, the temperature of the central part of the workpiece is always higher than temperatures of corners and edges and correspondingly, this setup is not fully controllable.

When the aluminum plate is replaced by a teflon plate, the temperature distribution is completely changed, such as shown in figure 5.18. Since teflon has a much lower thermal conductivity (0.25 $\mathrm{W/(m \cdot K)}$

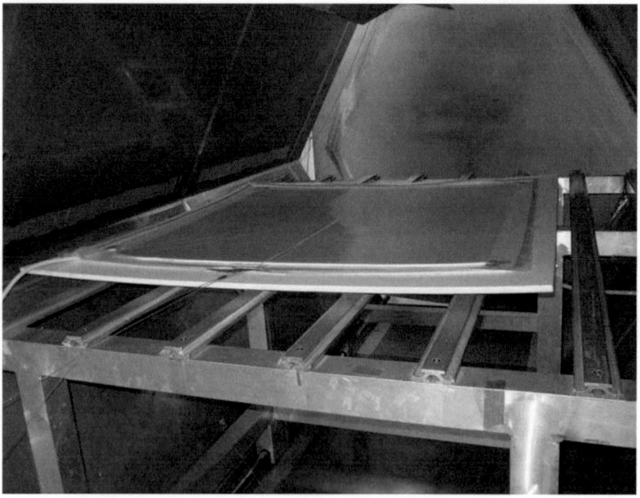

Figure 5.18. The setup with the teflon plate.

[tef]) compared with aluminum (237 $\mathrm{W/(m \cdot K)}$), the heat conduction effect caused by the teflon tool is weak. The temperature distribution of the teflon plate is similar to the temperature distribution of the workpiece. The temperature differences between different parts of the workpiece and their contacting parts of the teflon plate are simi-

lar, which lead to the same level of heat dissipation rates at different parts of the workpiece. Therefore the hot spots in the central part of the workpiece would be eliminated, which has been confirmed by the experimental results such as shown in figure 5.19.

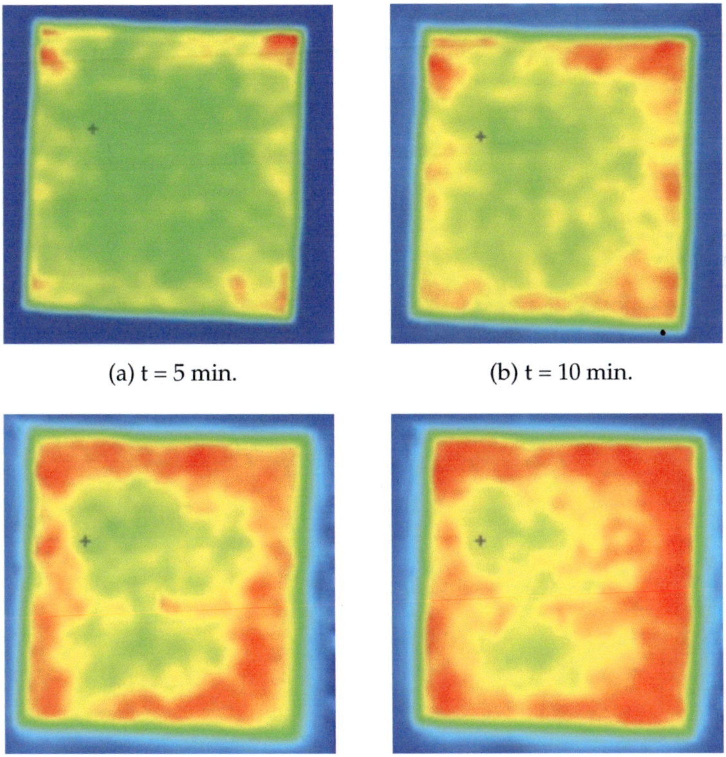

(a) t = 5 min. (b) t = 10 min.

(c) t = 15 min. (d) t = 20 min.

Figure 5.19. Heating process (from (a) to (d)) with the setup of teflon plate.

Although the hot spots are removed, there is another problem occurred due to the heat dissipation from the teflon plate to the metal table. When the workpiece is heated to a high temperature, the temperature of the teflon plate is also high, which causes deformations on the edges (see figures 5.20). For the central part of the plate, it has direct contact with the metal table and the heat dissipation rate from the teflon plate to the metal table is high. But for the corners and the edges, due to the deformations there is no contact as well as heat exchange between the teflon plate and the metal table. The different heat dissipation effects between the central and other parts of the workpiece lead to the cold area in the central and hot areas in the corners and the edges. This phenomenon exists during the entire heating process, and it is also not compensatable using different control methods.

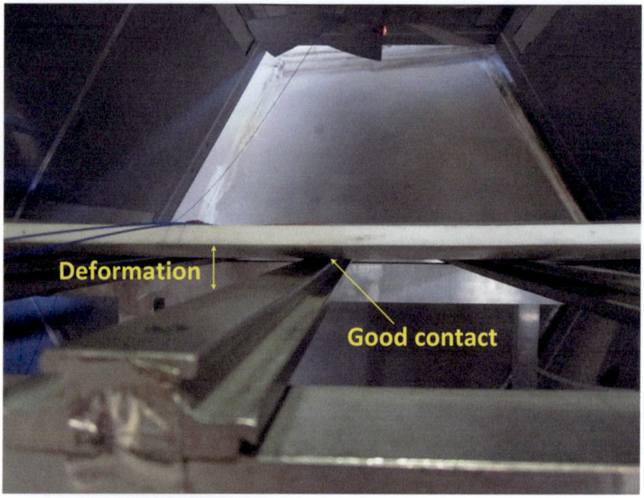

Figure 5.20. Deformation of the teflon plate (no contact between the plate and metal table).

Based on above observations, a simple and straightforward way to maintain a good controllability of the system is to reduce the size of the heated workpiece. The disturbance caused by different heat conduction effects of the tool is less severe for a small-tool setup, compared with the case in figure 5.17. As a result, the corresponding temperature distribution of the workpiece can be controlled using aforementioned control methods. Therefore a smaller setup, where the surface areas of both the workpiece and the aluminum plate are reduced into approximately $0.5\mathrm{m} \times 0.5\mathrm{m}$, is used in following experiments.

5.3.1. Performance of Adaptive Control Methods

In this part, four control methods have been tested and compared in the new HEPHAISTOS cavity 3 (new CA3, total 18 heating sources), including the PID control, the linear and the nonlinear MPC methods and the SPSA based NNC method. The PID control method is used as a benchmark for all other control methods. In the PID controller, only one temperature can be controlled and normally the maximum temperature is selected. The PID controller will calculate the control input value according to the following equation [ACL05]

$$
\begin{aligned}
u(k) &= K_\mathrm{p} \cdot e(k) + K_\mathrm{i} \cdot \sum_{i=0}^{k} e(i) + K_\mathrm{d} \cdot \frac{e(k) - e(k-1)}{\Delta t}, \\
e(k) &= Y_\mathrm{t}(k) - Y(k),
\end{aligned}
\tag{5.2}
$$

where $e(k)$ is the difference between the target temperature and the selected temperature value. The coefficient K_p, K_i, K_d are the proportional gain, the integral gain and the derivative gain, respectively. All control input element $u_i(k)$ in the control vector $\mathbf{U}(k)$ take the same value as $u(k)$.

The same setup as shown in figure 5.1 is used as the heated workpiece ($0.5\mathrm{m} \times 0.5\mathrm{m}$), and temperatures of five different locations are measured by the infrared camera and controlled, such as in figure 5.21.

The silicone rubber setup is used instead of real CFRP material in the experiments. In principle the control methods introduced in this dissertation can be applied to all different kinds of materials. For different materials, the control performance would be different, and the final controlled temperature distribution mainly depends on the thermal conductivity of the setup or material. The thermal conductivity directly determines the accuracy of the estimated model and the effective control diversity. From this point of view, the silicone rubber setup is a good experiment material, because it has a similar thermal conductivity with certain CFRP materials, e.g. the thermal conductivity of silicone rubber is $0.2\mathrm{W}/(\mathrm{m} \cdot \mathrm{K}) \sim 1.3\mathrm{W}/(\mathrm{m} \cdot \mathrm{K})$ [Sil12], and the thermal conductivity of a typical CFRP epoxy prepreg is $0.6\mathrm{W}/(\mathrm{m} \cdot \mathrm{K})$ [AD10]. If the control methods can perform well using the silicone rubber setup, it is reasonable to believe they can achieve a similar temperature homogeneity using the real CFRP prepregs.

The target temperature curve defined in the experiments is shown by figure 5.22. Considering the safety issue and the reliability of the vacuum bagging, the maximum target temperature is set to be 100°C and the temperature increasing rate is 8°C/min. The target temperature curve is defined as shown in figure 5.22, which is similar with the target temperature curve used in real CFRP curing process [PPW+02].

A typical control performance of PID is shown in figure 5.23. The control period used in the PID controller is $\Delta t = 1.5$ s. In figure 5.23, the first temperature T1 is used as the controlled temperature. For the controlled temperature T1, it can be perfectly controlled by the PID controller and it follows the target temperature curve closely during the entire heating process. However, the problem for the PID controller is that all other temperatures are not controlled at all. All different temperature curves have the same changing trend and the temperature differences between each other are almost constant during the flat-temperature period. Although the final temperature window ΔT_1 and ΔT_2 (temperature difference between the maximum and the minimum temperatures) in the end of both flat-temperature periods

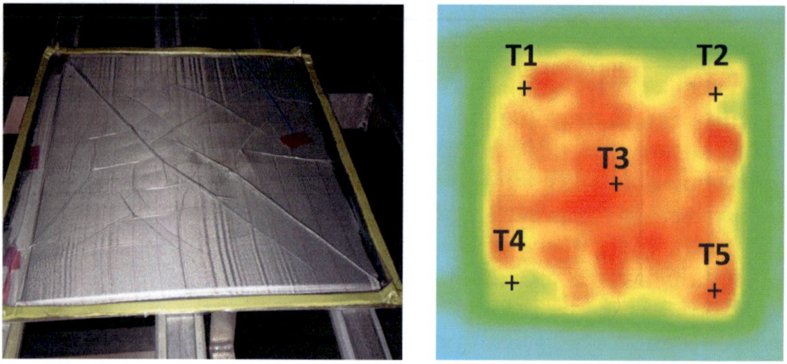

(a) Workpiece used in the experiment. (b) Measured temperatures.

Figure 5.21. Picture of the workpiece and measured temperatures.

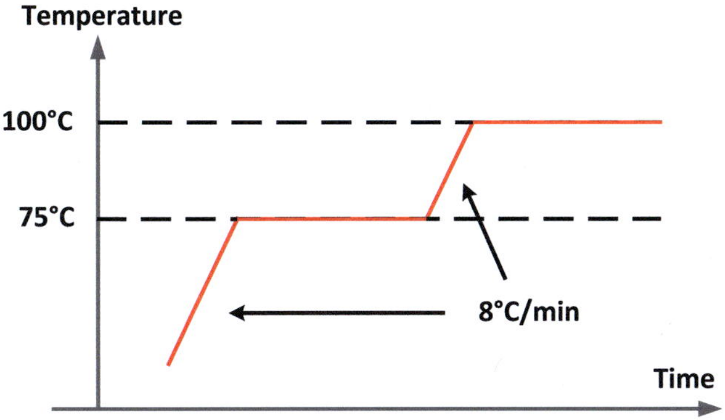

Figure 5.22. Illustration of the target temperature curve.

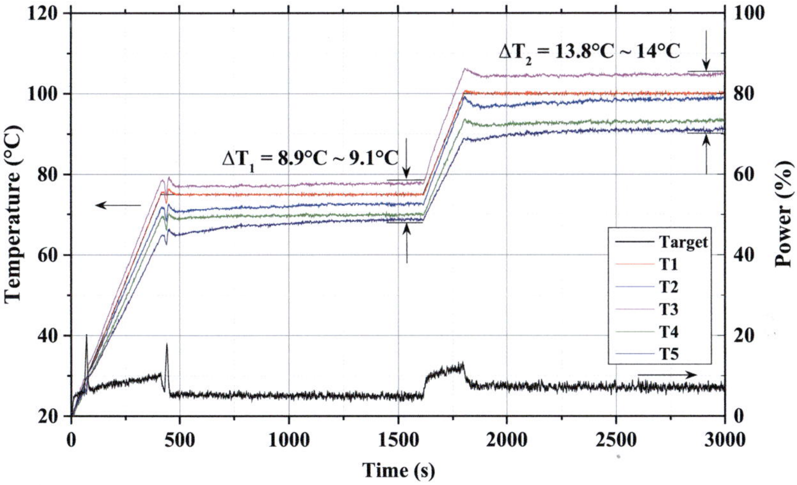

Figure 5.23. Control performance of PID control method (new cavity 3, 12 sources).

are slightly smaller than the value in the beginning, that is due to the combined thermal conduction effect of the aluminum plate and the workpiece itself. The PID controller itself does not have the ability to reduce the temperature window and improve the temperature distribution. It should be noted that the second temperature window ΔT_2

is significantly larger than ΔT_1. It indicates that the temperature window in PID control mainly depend on the setup and other external factors (such as the target temperature), which also reflect the fact that the PID controller does not have the active ability to improve the temperature distribution.

In principle, all 18 heating sources within the new CA3 could be used for the controlling, but in the above as well as following experiments, only 12 of them (the first two modules) are used. That is because for the other 6 heating sources using the low-ripple power supplies, they are not able to output power lower than 10% of the maximum power. They can work properly when the control input value $u(k)$ is larger than 0.1, but when the value is lower than 0.1, the practical output power is always fixed at $10\% \times 2\text{kW} = 200\text{W}$. This behavior would cause extra disturbances to the system estimation and it is also not reliable considering some safety issues. Therefore in these experiments, only the first 12 heating sources are used. In the future, additional micro controllers or frequency converters will be attached to the low-ripple power supplies to make sure that they will work in a securer PWM mode.

However, for PID the number of heating sources and different types of power supplies do not have any severe impacts to the control performance. The control results of the PID controller in the old HEPHAISTOS CA3 using 36 heating sources are shown in figure 5.24. Comparing the results in these two experiments, they have the similar control behaviors and the same level of final temperature windows. More heating sources does not necessarily lead to a better control performance and a more homogeneous temperature distribution. As long as the number of the heating sources is enough to provide a diverse heating (which depends on the specific application) and they are properly distributed spatially, the control diversity as well as the corresponding control performance can be guaranteed. This phenomenon is also reflected by some of the following results, and it is one of the reasons why in the new CA3 the number of sources is halved.

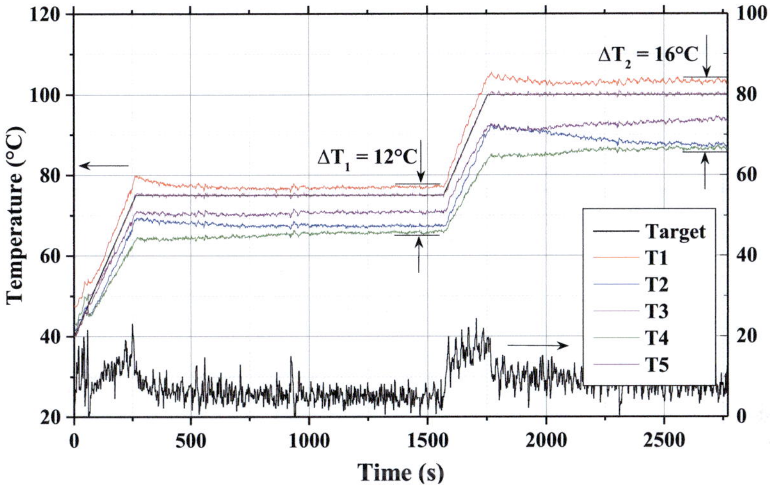

Figure 5.24. Control performance of PID control method (old cavity 3, 36 sources).

Model Predictive Control

Besides the PID control method, all other adaptive control methods introduced in this dissertation have also been tested. In the tests of the MPC methods, the control period is $\Delta t = 1.5$ s and the prediction length is $p = 5$. In theory, the prediction length should be defined as large as possible, because a large prediction length normally leads to a better control performance [Wan09] [CA13]. But in practice, a large prediction length is not affordable because of the increased computation time, especially for the nonlinear MPC. In order to keep a fast control pace (1.5 s), the prediction length is set to be 5 in our experiments. The performance of the linear and the nonlinear MPC methods are shown in figures 5.25.

The two dashed lines in figure 5.25a are used to denote the control 'accident' occurred in the heating process, such as in figure 5.26. In HEPHAISTOS, to guarantee the reliability of the whole system, a external watchdog timer [Mur00] is equipped as a hardware switch to the power supplies. During the normal operation, the LabVIEW con-

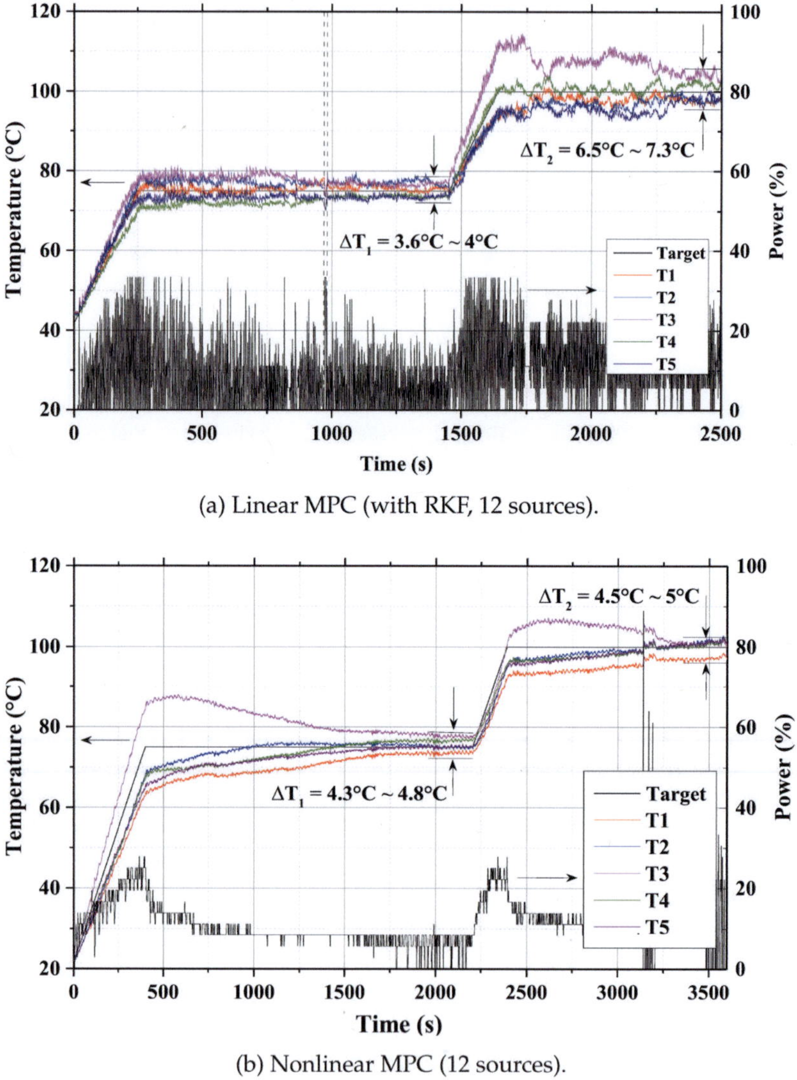

(a) Linear MPC (with RKF, 12 sources).

(b) Nonlinear MPC (12 sources).

Figure 5.25. Control results of the linear MPC (a) and nonlinear MPC (b).

trol interface regularly sends a pulse signal to the watchdog timer e 200 ms, to prevent it from elapsing. If any software or hardware accidents happen, the LabVIEW control interface fails to the send this pulse signal. Then the watchdog timer will elapse and switch off the power supplies.

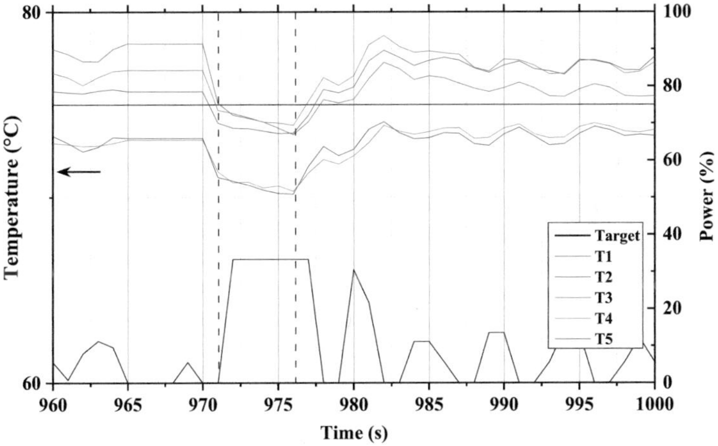

Figure 5.26. Control 'accident' in the heating process.

In practical applications, since LabVIEW is not a pure real-time software, the time cost in each LabVIEW loop is not fixed. In some cases the loop takes more than 200 ms, and then the watchdog timer will elapse and switch off the power supplies. What happens in the next is shown in figure 5.26. No matter what control input is sent to the HEPHAISTOS cavity, there is no microwave heating power and hence the temperatures drop quickly. In such cases, the LabVIEW control interface and the MATLAB control program are not interrupted since there is no signal from the power supplies or the watchdog timer to report their status. They continuously update the estimated system model and calculate new control input, based on the incorrect power-temperature pairs.

This kind of failure is difficult to be eliminated from a hardware aspect (the timeout value of the watchdog timer is fixed), and it is also hard to be detected in the control program. Normally when a single 'accident' occurs, the system can recover from it quickly, owing to the adaptive system identification algorithms. However, when several accidents occur in a row during a short time, it could cause much larger unexpected system estimation errors and corresponding wrong control inputs (see the results later)

From the results in figure 5.25, it is clear that both the linear and the nonlinear MPC methods are much better than the PID control method, with respect to the final temperature window. Both of them have much smaller final temperature windows compared with the PID controller. More important, the temperature curves in figures 5.25 have obvious converging behaviors, which indicate temperatures of different locations are effectively controlled towards the target temperature value. For example, in the first figure 5.25a, the temperature window in the beginning of the flat-temperature period is more than 10°C. During the heating process, all different temperature curves are gradually converging to the target temperature value (75°C) and the temperature window is also decreasing. In the end of the flat-temperature period, the final temperature window ΔT_1 is between 3.6°C and 4°C. This converging behavior occurs in both the linear and the nonlinear MPC methods, and it reflects the active temperature distribution improving ability of both MPC methods.

For the linear MPC, there are two system identification methods available such as introduced in chapter 3. Both RLS and RKF have been tested and compared in our experiments, and corresponding results can be found in figures 5.27. There is no obvious difference between the performance of RLS and RKF, which also coincides with the system identification results shown before. It should also be noted that the control performance of the linear MPC method with more heating sources (in figure 5.27) is comparable to the results in the new CA3 with fewer heating sources(in figure 5.25a). It indicates that the number of heating sources does not influence the control performance and final temperature homogeneity of the linear MPC method, which is similar with the situation in the PID control.

The control performance of nonlinear MPC is similar with that of the linear MPC. The converging behavior is obvious and the final temperature window is also much smaller than in the PID controller. A problem occurred in the nonlinear MPC is that the large temperature overshoot in the beginning of the flat-temperature period (see figure

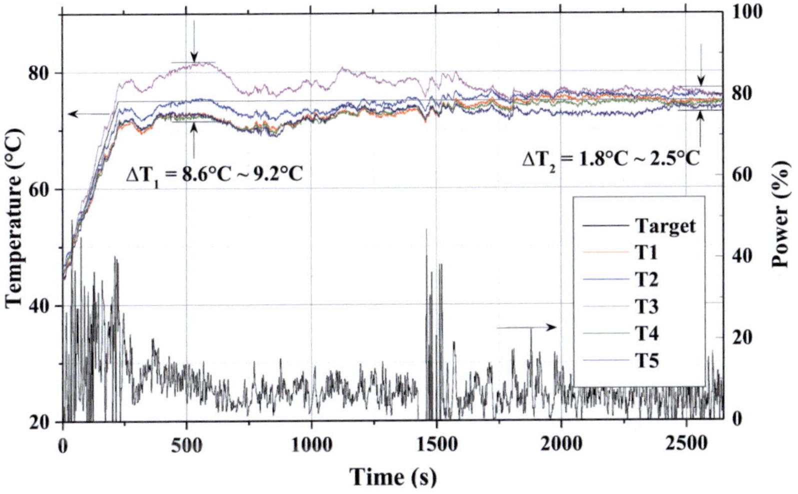

(a) Linear MPC using RLS (old CA3, 36 sources).

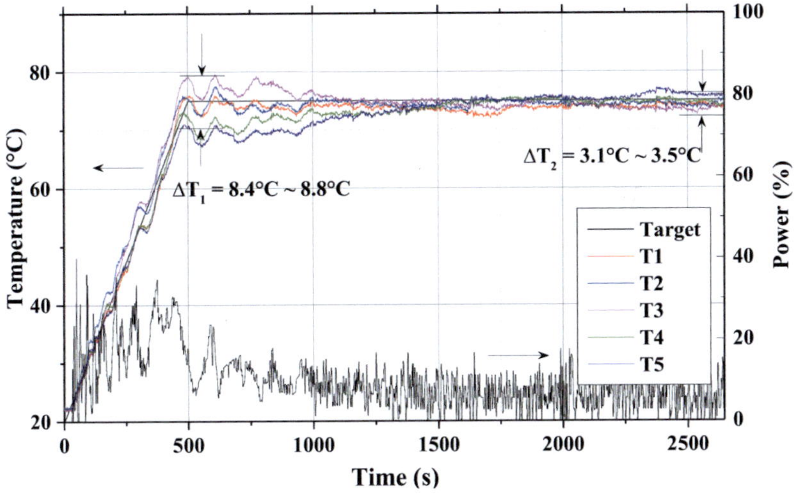

(b) Linear MPC using RKF (old CA3, 36 sources).

Figure 5.27. Control results of Linear MPC with RLS (a) and RKF (b).

5.25b). The problem is caused by the error from either the system identification or the GA controller. The high control input dimension of HEPHAISTOS is a large computation burden for both of them, especially for the GA algorithm. The overall searching space in the GA algorithm increases exponentially with the input dimension. When all other parameters (such as the number of generations used and the number of the individuals in each generation) are fixed, a higher input dimension leads to a much larger searching space in the GA algorithm, which correspondingly lowers the probability that a good control solution can be obtained. In practical experiments, the most simple and effective way to reduce the temperature overshoot and enhance the entire control performance is to use less heating sources.

For example, in the experiment shown by figure 5.28, all heating sources below the metal table are not used during the heating process (in each module only sources No. 2, No. 4, No. 10 and No. 11 are switched on, see figure 2.13). Overall 8 heating sources are used for the controlling. On the one hand, the computation burdens for both the system estimator and the controller are largely reduced. On the other hand, the total 8 sources still guarantee a huge control space and control diversity. With fewer heating sources, the accuracies of the system identification and the GA controller both can be improved. Hence, from the results in figure 5.28, it is clear that both the temperature overshoot and the final temperature window are significantly reduced using less sources.

As aforementioned in the system modeling part (chapter 2) and the beginning of this section, thermal conduction plays an important rule in the model derivation as well as controlling process. In order to further test how different conduction effects influence the control performance, a different setup was implemented in our experiments, such as in figure 5.29. This setup has the same aluminum plate, vacuum bagging and other components, except one big difference. Instead of one big silicone rubber workpiece, in the second setup 5.29 there are five independent small workpieces. Compared with the first setup 5.21, apparently the second setup has a much lower influence from the thermal conduction and convection parts. In other words, all cooling effects including the thermal convection, radiation and conduction for the five workpieces are almost the same. Therefore in such cases, the properties of each control method can be more greatly reflected.

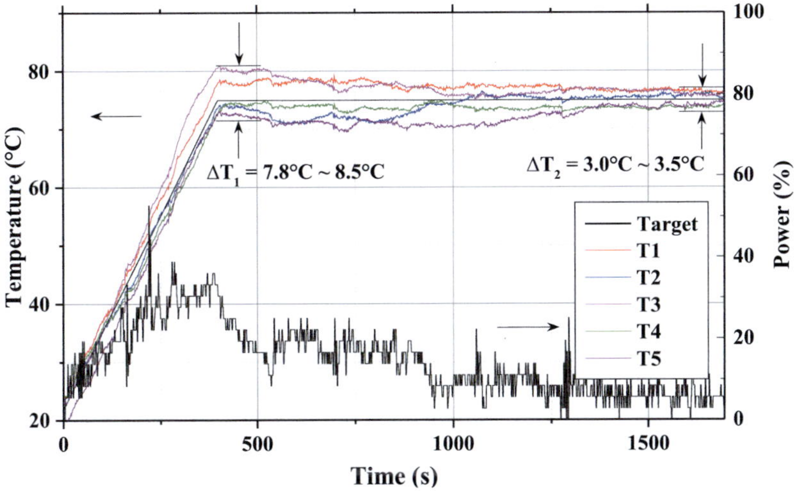

Figure 5.28. Control results of nonlinear MPC using 8 heating sources (new CA3).

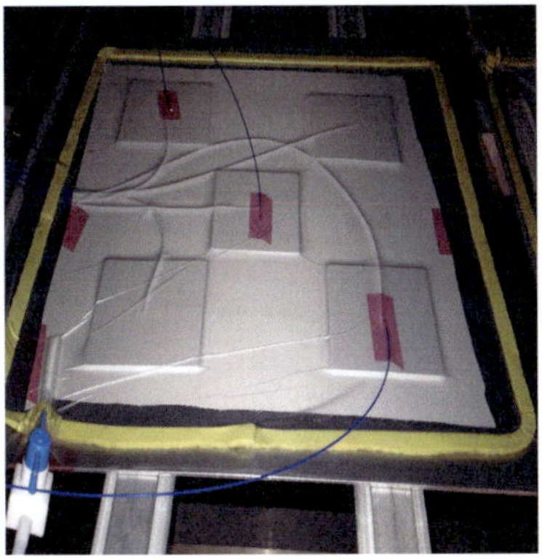

(a) The second setup used in the experiment.

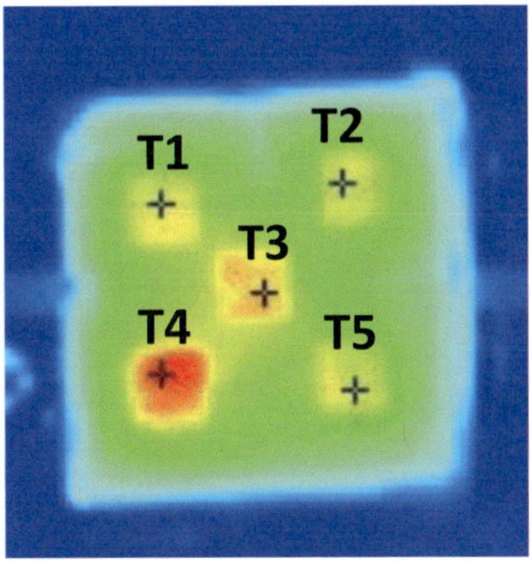

(b) Thermal view and the temperatures measured.

Figure 5.29. The setup with independent small workpieces.

The control performance of the three aforementioned control methods is presented in figures 5.30.

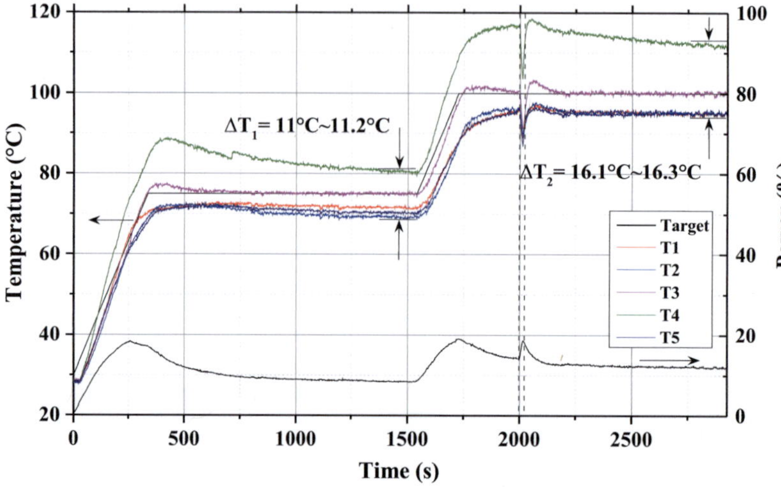

(a) PID (new CA3, 12 sources).

Comparing the results in figures 5.30 to the results of the first setup (in figures 5.23 and 5.25), there are a number of obvious distinctions. For the PID controller, the temperature windows are even larger than the results in the first setup, which indicates that the conduction based temperature improvement effect in the second setup is much lower than in the first setup. Unlike the PID controller, the performance of the linear and the nonlinear MPC methods is both better than the results in the first setup (see figure 5.25), regarding the temperature overshoot and the final temperature window aspects. For the linear MPC, it has smaller temperature windows as well as much lower temperature overshoot and oscillations, especially in the high temperature range. This performance fully reflect that without extra disturbances from the thermal conduction, the model constructed in the equation 3.41 is more accurate and close to describe the heating dynamics, which results in better control results.

For the nonlinear MPC, the performance improvement is even more obvious than the linear MPC. That is partly because less heating sources are used. At the same time, it profits more from the reduced impact of the thermal conduction. In both MPC methods, the influences from all cooling effects including the thermal conduction can be

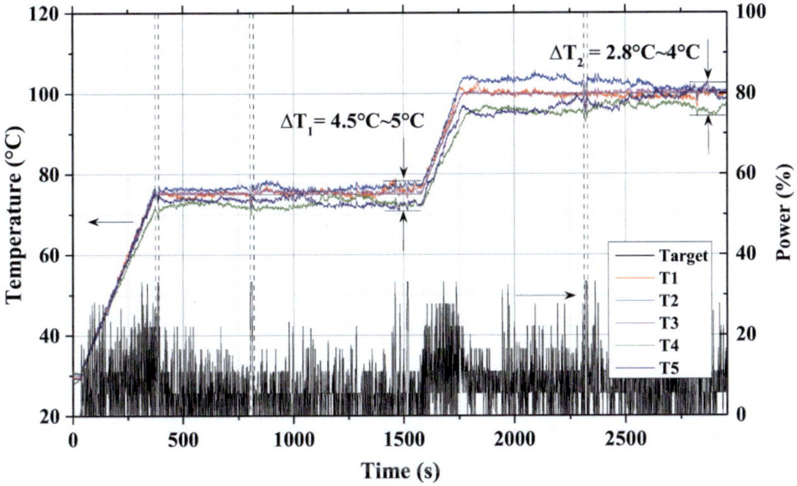

(b) Linear MPC (new CA3, 12 sources)

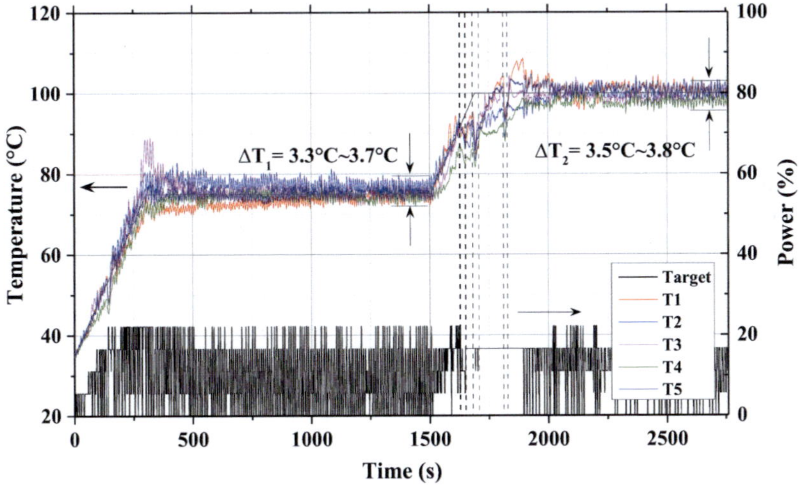

(c) Nonlinear MPC (new CA3, 8 sources).

adjusted by the self-cooling matrix **A**. But due to the system identification procedures introduced in chapter 3, the nonlinear MPC method has much less opportunities to update the matrix **A**. Therefore it does

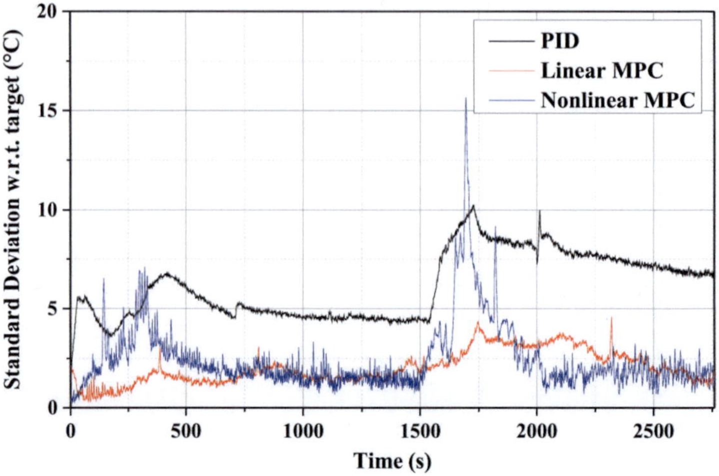

(d) Standard deviation with respect to the target temperature.

Figure 5.30. Control performance of PID, linear MPC and nonlinear MPC in the second setup. Due to slightly different duration in the three experiments, the comparison of standard deviation lasts until 2760 s.

not have the flexibility to deal with fast varying effects from the thermal conduction, and this kind of influence is further transmitted to the estimation error of the heating effect. Using the setup 5.29, the cooling effect of each workpiece is more stable than the first setup, and correspondingly the accuracy of the nonlinear system identification is higher than the case with the first setup. Consequently, the final control performance is improved.

In figure 5.30c, when the target temperature is increasing from 75°C to 100°C, there are three control accidents occurred consecutively. Due to these three accidents, the controlled temperatures fail to follow the target temperature curve and the standard deviation becomes much larger than the normal level (from 1600 s to 1800 s). But after that period, the system estimator takes a short time to recover from the accidents and the control behavior is back to normal quickly. It also reflects the robustness of the system identification algorithm and the MPC method.

A phenomenon is worthy noted that in former results (figures 5.23 and 5.25), the highest temperature is mostly measured in the central part of the workpiece (T3). But in the second setup, the hot spot is possible to occur in any workpieces. For example, in the experiment shown in figure 5.30a the highest temperature is always T4. This phenomenon denotes the same cooling effects of these five workpieces and also confirms that the active temperature distribution improvement is generated by the MPC methods.

Neural Network based Control

The SPSA based NNC method has also been tested using the same setup and temperature measurement scheme shown in figure 5.21. Before its real implementation, it was firstly tested in a number of simulations. These simulations use the similar idea that is the system identification part, which is that a NN was trained based on real experimental data to act as the plant, and then the NNC was applied to control this well-trained NN. The structure of simulations is shown in figure 5.31.

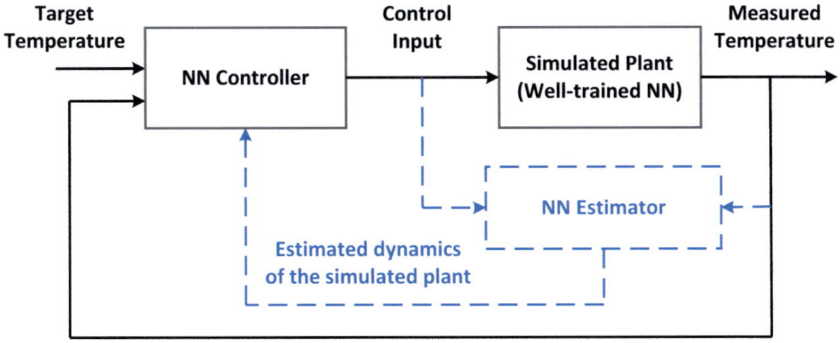

Figure 5.31. Structure used in the NNC simulations (the NN estimator was trained based on experimental data from the new CA3 with 8 heating sources).

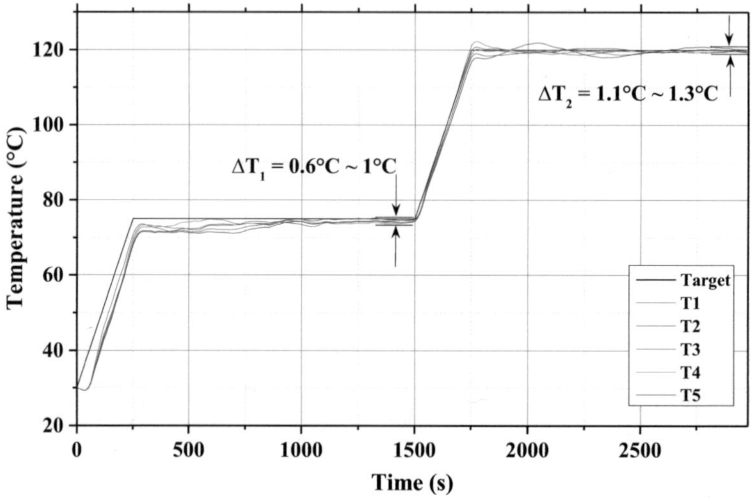

Figure 5.32. Simulation results of NNC.

The results from the simulations are shown in figure 5.32. From the simulated results, it is clear that the simulated plant is well controlled using the NNC method and the final temperature window is small. Comparing with former presented MPC results, there are one notable difference between the NNC and MPC methods. It is that in the beginning of the control process, the controlled temperatures always lag behind the target temperature. Because the controller has to take a certain time to update its weights from the plant or the NN estimator. After the update is finished, the controlled temperatures gradually converge to the target temperature in a faster pace and they track the target temperature perfectly.

The same control behavior also occurred in the real experiments of NNC, as shown by figure 5.33. At both raising-temperature periods, the controlled temperatures have lower temperature increasing rates than the required value. But on the other hand, the lower heating rates lead to a small temperature window in NNC during the whole heating process, and there is no temperature overshoot at the flat-temperature period. This is the advantage that NNC outperforms the MPC methods. Compared with the PID controller, NNC is also much better because of the better temperature homogeneity.

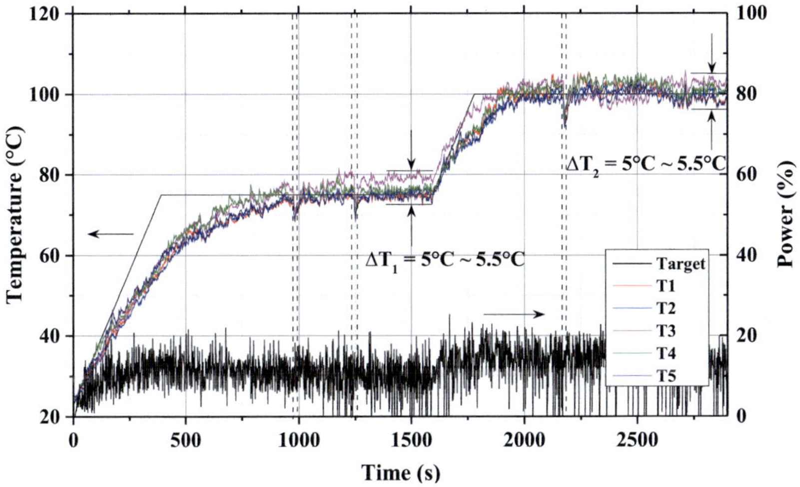

Figure 5.33. Control performance of NNC (new CA3, 8 sources).

When the weights of the NN controller are properly initialized, this temperature lag phenomenon can be significantly reduced. For example, the control results of another NNC experiment is shown in figure 5.34. In this experiment, the controller is initialized with the same weights as the trained controller in 5.33. In other words, after the experiment shown in figure 5.33, the same controller is used again in the same environment to deal with the same control task.

From the results in figure 5.34, a big improvement can be observed is that the controlled temperatures follow the target temperature closely almost from the beginning of the control process. There is no temperature overshoot and the temperature window is smaller than the results shown in figure 5.33. This performance improvement is due to the appropriate initialization of the NN controller. A good initialization would speed up the entire controller update and lead to a more reliable control performance in the beginning period.

From the control results shown by these pictures, all three adaptive control methods, including the linear MPC, the nonlinear MPC and

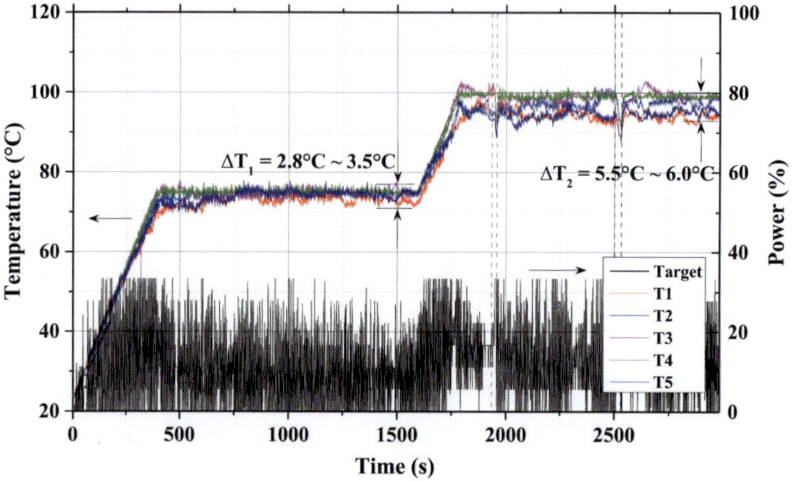

Figure 5.34. Control results of NNC initialized with the same weights of the controller trained in 5.33 (new CA3, 8 sources).

the NNC methods, achieve the same level of control performance, regarding the final temperature window and the temperature homogeneity aspects. Although there is no essential difference among the control results, different properties can be observed for individual control methods. The linear MPC method is the most robust and easily applicable control method. Benefiting from the well established system identification method (which estimates **A** and **B** simultaneously) and (partially analytical) control solution, the computation requirements in the linear MPC is the lowest among the three methods. Therefore it is able to control high input/output dimensional systems with a guaranteed control performance. Compared with the linear MPC, the performance of the nonlinear MPC is partly influenced by the input dimension. In order to get a more reliable control performance, the number of heating sources has to be limited. However, on the other hand, as long as the control diversity and the performance can be guaranteed, there is no need to use more heating sources. In addition, the nonlinear MPC could achieve a slightly better control result than the linear MPC method in the second setup, which means it is preferred to be applied in situations with more stable cooling effects.

In both MPC methods, only the parameters in the system estimator have to be updated. In other words, as long as the system estimator is able to estimate and predict the dynamics of the plant perfectly, the corresponding control solution is also perfect or guaranteed. But in the NNC system, both the controller and the system estimator have to be updated and trained during the heating process. Since the controller is a NN trained by the unsupervised learning scheme, it takes much longer time to reach the same level of control performance as the MPC methods. Nevertheless, a well trained NNC can achieve even better control performance than the MPC methods, without any extra requirements of the setup or the material. There is also another big advantage that the NNC method outperforms the MPC methods, which is that the NNC method is model-free. It can be implemented to control both linear and nonlinear plants and no additional physical insight is needed. Therefore the NNC method could be more easily transferred to any other control applications.

5.3.2. Performance of Intelligent Control Methods

Compared with the adaptive control methods, the implementation of the intelligent control method is more complicated. Additional LabVIEW operations are required to obtain the coordinates of the maximum and the minimum temperatures such as illustrated in figure 5.35. These coordinates are then compared with predefined region boards (such as in figure 4.13) to determine which region the highest and the lowest temperatures are.

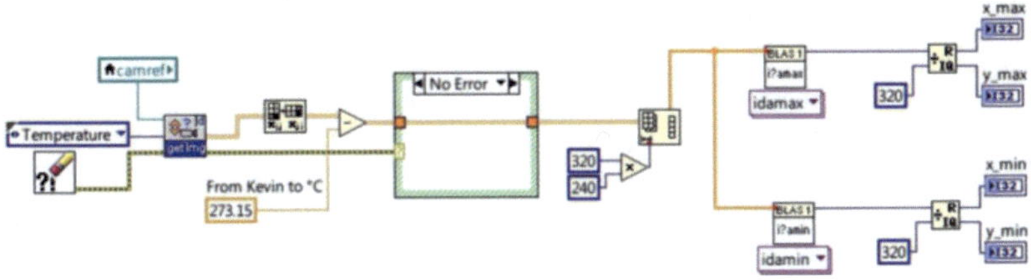

Figure 5.35. LabVIEW program (steps) to obtain the coordinates of the maximum temperature.

As mentioned in chapter 4, the $Q(\lambda)$ based RLC is only tested in the flat-temperature period. The setup 5.21 is used in these experiments and the target temperature curve is the same as in figure 4.11. Before the controller is tested in real experiments, it was also tested in a number of simulations. The control task in the simulations is different to the task 4.30 mentioned in chapter 4 because there is no way to simulate the whole temperature distribution. The same well-trained NN used in the NNC simulations 5.31 is also used here. Instead of the maximum and the minimum temperatures of the whole thermal picture, the maximum and the minimum temperatures from the five simulated temperatures are used as the controlled temperatures. Correspondingly, the indexes of these two temperatures are considered as the location states. For instance, T1 is the highest temperature is equivalent to the idea that the hot spot is the region R1 (illustrated in figure 4.13).

This simulation can be considered as the simplified version of the real control task. The purpose of this simulation is to test if all five temperatures can be limited within a certain range by only controlling the maximum and the minimum temperatures. The corresponding simulation results are shown in figure 5.36. The entire simulation lasts 9000 s, but the controller takes around 3000 s (3000 control periods) to reduce the temperature window from more than 10°C to about 1°C. This result reflects that the $Q(\lambda)$ RLC method is effective to affect all other uncontrolled temperatures by only controlling the maximum and the minimum temperatures.

Based on this result, the hybrid control structure (figure 4.14) was tested in real experiments. The total controlled area is slightly smaller than the one shown in figure 4.13 to prevent any uncontrollable locations in the corners or edges. The corresponding control results are shown in figure 5.37.

In the first trial 5.37a, the raising-temperature period is controlled by using the linear MPC controller. In the flat-temperature period, the temperature window between the maximum and minimum temperatures is not effectively reduced. Because the $Q(\lambda)$ controller has to learn the correct state-action values during the controlling and it also has to take many exploration control actions. The exploration probability ϵ used in the $Q(\lambda)$ controller decreases from 0.9 in the beginning

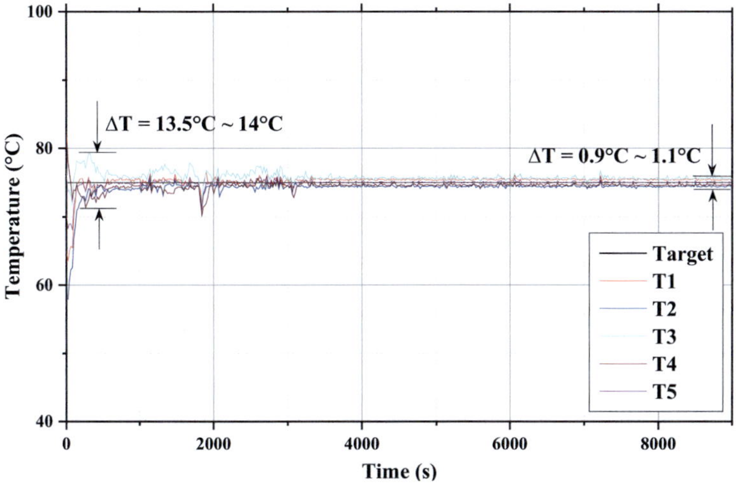

Figure 5.36. Simulation results of the $Q(\lambda)$ based RLC method.

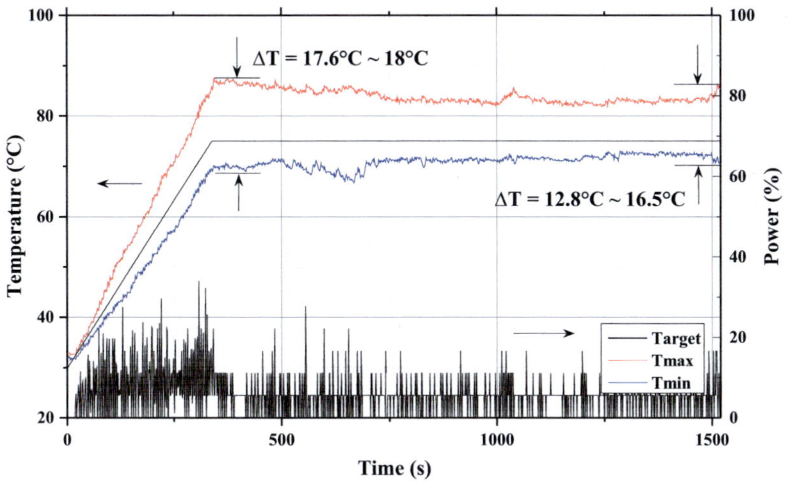

(a) The first trial.

to 0.1 in the end of the heating process. In order to have a more stable control performance, meanwhile also keep a certain exploration ability, the exploration probability used the in the last two trials is set to be 0.1.

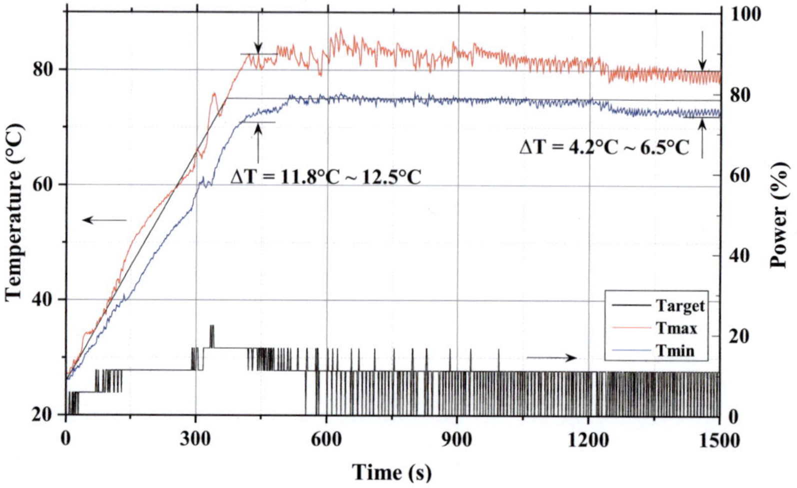

(b) The second trial (using the same controller trained from the first trial).

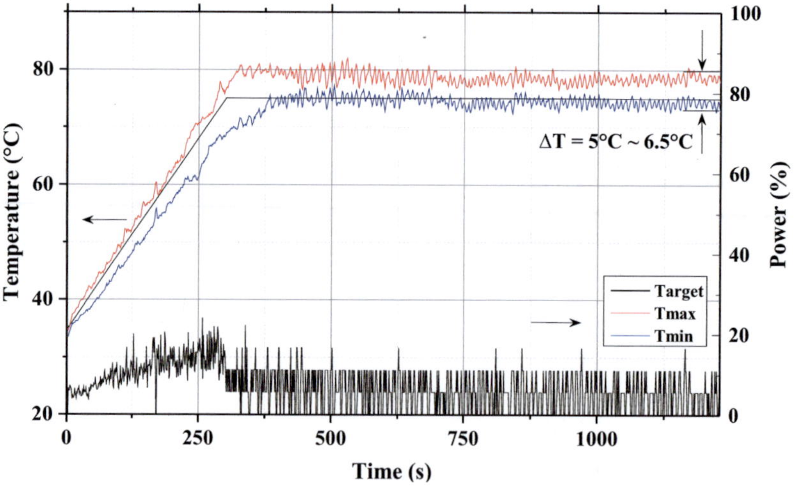

(c) The third trial (using the same controller trained from the second trial).

Figure 5.37. Control results of the $Q(\lambda)$ based RLC method in three consecutive trials.

After the first trial, the preliminarily trained $Q(\lambda)$ is directly used in the second trial, just as the case in the NNC experiments. The first raising-temperature period of the second trial is controlled using the nonlinear MPC method. In the flat-temperature period, the minimum temperature is oscillating around the target temperature, therefore according to the principle defined in the hybrid control scheme (figure 4.14), the activated controller switching between the 'no power' control and the $Q(\lambda)$ controller. The temperature state mainly stays as $(S_{\max} = 0, S_{\min} = 1)$, which means the minimum temperature is in the desired temperature range but the maximum temperature is not. Although the locations states may vary during this oscillation period, all state-action values related with $(S_{\max} = 0, S_{\min} = 1)$ are tested for multiple times and updated to a higher accuracy than the others. After the oscillation, the states of the system enters into the desired temperature range with $(S_{\max} = 1, S_{\min} = 1)$, which means all temperatures in the controlled area are within the desired range and the control target is achieved.

In order to further test the effectiveness of the $Q(\lambda)$ controller, the third trial is carried out consecutively. The state-action values trained in the second trial is loaded in the third trial. The raising-temperature period of this trial is controlled by a NN controller. On the one hand, the results in figure 5.37c are similar with the results in the second trial. Both the maximum and the minimum temperatures stay in the desired range for the most time of this trial, which means the $Q(\lambda)$ controller is effective to control the temperature distribution. On the other hand, its effectiveness can not be fully confirmed by the results in figure 5.37c. Because the temperature state in the third trial is the same as the state in the second trial. In other words, after the update process in the second trial, the $Q(\lambda)$ controller may already have accurate state-action values for all state-action pairs related to the state $(S_{\max} = 0, S_{\min} = 1)$. But it does not mean this $Q(\lambda)$ controller also have the capability to deal with other temperature states. For example, if the temperature state in the third trial is $(S_{\max} = 1, S_{\min} = 0)$ (the maximum temperature is within the range but the minimum temperature is not), the controller may need a long time to update the state-action values related with this state again.

Despite the question whether the $Q(\lambda)$ controller is fully trained during the experiments, the results in figures 5.37 show its great potential and capability to control the entire temperature distribution of the workpiece. Unlike the adaptive control scheme that is based on the individual temperatures, the $Q(\lambda)$ based RL controller is able to directly control the maximum and the minimum temperatures. The entire temperature distribution is guaranteed to improve if both these two temperatures are limited. From this point of view, the intelligent RLC scheme is more usefull for practical industrial applications.

6. Summary and Conclusion

This dissertation gives a detailed introduction of microwave heating systems with spatially distributed heating sources. The innovative temperature control methods that are implemented in HEPHAISTOS achieved promising performance with respect to the temperature homogeneity of the heat load. The basic principles of microwave heating and microwave heating systems are explained. In order to accurately describe the dynamics of the the microwave heating process, three different heating models have been developed: the linear heating model, the nonlinear heating model and the neural network based black-box model. Parameters within each model are estimated online using the corresponding system identification method.

Based on the estimated parameters, various adaptive control methods, including the model predictive control (MPC) method and the neural network based control (NNC) method, are implemented. Both control methods have been modified and optimized according to the practical situations and requirements of the microwave heating system, in order to improve control performance.

Besides the adaptive control scheme that focuses on individual measured temperatures, another reinforcement learning based intelligent control (RLC) scheme is built in this dissertation. The dynamics of HEPHAISTOS are described by a Markov decision process. Then the control task is solved using a look-up table based $Q(\lambda)$ controller, which is simplified and optimized based on the empirical knowledge.

The aforementioned heating models and control methods have been tested in practical experiments and the corresponding experimental data are presented. Firstly the validity of different models is verified through numerous preliminary tests. Then different system identification algorithms are tested and compared using real experimental data.

Finally, all aforementioned control methods are tested in practical experiments, and different properties have been observed for individual control methods.

For the adaptive control scheme, both MPC and the NNC are able to let multiple temperatures converge at the target temperature. The final temperature distribution and the temperature window in both control methods are significantly improved compared with the conventional PID controller. The control models used in the MPC method imply that they are well suited to materials and setups with lower thermal conductivities or multiple independent workpieces. The NNC method can be used in more general situations without special requirements on the properties of the heated material.

For the adaptive control scheme, the intelligent $Q(\lambda)$ based reinforcement learning controller also shows great control performance, reflected by its effective control of the temperature window between the maximum and the minimum temperatures. It is suitable for applications where the entire temperature profile can be monitored in real-time. Compared with conventional control schemes, it is more powerful because the temperature distribution can be directly controlled and improved.

The main contributions of this dissertation are:

- The nonlinear state-space and the neural network microwave heating models developed in this dissertation provide powerful alternatives to the traditional linear microwave heating models (as in [HPE97] and [RCVI99]). The dynamics of the distributed microwave heating systems are accurately described and estimated by these multiple-input multiple-output (MIMO) models that have been constructed in this dissertation.

- Based on the MIMO models, different MIMO control systems are designed and applied to control the distributed microwave heating system - HEPHAISTOS. The temperature homogeneity (reflected by the final temperature window ΔT) obtained using the the advanced MIMO control system ($\Delta T = 4°\text{C}$ - $6°\text{C}$ at $100°\text{C}$ for both MPC and NNC) is much better than the conventional PID controller ($\Delta T \geq 13°\text{C}$ at $100°\text{C}$). The experimental results presented in this dissertation provide evidence that the heating performance of the distributed microwave heating system can be significantly improved by the advanced MIMO control system. The global temperature window (temperature window of the entire heat load) achieved by the RLC system (5°C - 6.5°C) is also comparable to normal commercial industrial heating equipments, such as the standard in [Ind13](±2.5°C).

- The control concept that combines the real time thermal picture and the intelligent control system ($Q(\lambda)$ controller) is unique. It is able to control the complete temperature distribution of the entire workpiece, and more straightforward than conventional methods based on individual temperature values. Its effectiveness has been proved by experimental results.
- The temperature control system and software developed in this dissertation have a great potential to be implemented in practical industrial applications, to promote the use of microwave heating to broader areas. They can also be transferred to other distributed microwave feeding systems, such as the octagonal microwave cavity (proposed in [LLHG14]) and other related applications.

Based on the work done in this dissertation, the MIMO temperature control system could be further optimized in order to improve its reliability and control performance. For example, more experiment parameters could be monitored and taken into account in the modeling process, e.g. the air flow speed in the oven and the real-time temperature of the metal table. The involvement of such parameters will increase the accuracy of the control model, as well as the reusability of the controller. With a high reusability, control performance could be improved by reusing the former well-trained controller (such as the results of repeated trials shown in figures 5.34 and 5.37), especially for the reinforcement learning controller. In this case, a more powerful reinforcement learning controller could be built based on a complicated function approximator (nonlinear function or neural network) and long time practical experiment training. Another research direction would be the combination of microwave and hot air heating. The hot air flow could be controlled using an independent PID controller to heat the cold area of the load, which will be of great help to achieve a more homogeneous temperature distribution between the inner core and surface of the entire load.

Bibliography

[ACL05] Kiam Heong Ang, Gregory Chong, and Yun Li. PID control system analysis, design, and technology. *Control Systems Technology, IEEE Transactions on*, 13(4):559–576, 2005.

[AD10] 3M Industrial Adhesives and Tapes Division. Technical data from 3M. Technical report, 3M Company, 2010.

[ADDG92] KG Ayappa, HT Davis, EA Davis, and J Gordon. Two-dimensional finite element analysis of microwave heating. *AIChE Journal*, 38(10):1577–1592, 1992.

[ÅJPJ08] Bernt M Åkesson, John Bagterp Jørgensen, Niels Kjølstad Poulsen, and Sten Bay Jørgensen. A generalized autocovariance least-squares method for Kalman filter tuning. *Journal of Process control*, 18(7):769–779, 2008.

[ALL+09] John Asmuth, Lihong Li, Michael L Littman, Ali Nouri, and David Wingate. A Bayesian sampling approach to exploration in reinforcement learning. In *Proceedings of the Twenty-Fifth Conference on Uncertainty in Artificial Intelligence*, pages 19–26. AUAI Press, 2009.

[Alp04] Ethem Alpaydin. *Introduction to machine learning*. MIT press, 2004.

[AM07] Brian DO Anderson and John B Moore. *Optimal control: linear quadratic methods*. Courier Corporation, 2007.

[AMI02] Vijanth Sagayan Asirvadam, SF McLoone, and GW Irwin. Parallel and separable recursive Levenberg-Marquardt training algorithm. In *Neural Networks for Signal Processing, 2002. Proceedings of the 2002 12th IEEE Workshop on*, pages 129–138. IEEE, 2002.

[AN05] Pieter Abbeel and Andrew Y Ng. Exploration and apprenticeship learning in reinforcement learning. In *Proceedings of the 22nd international conference on Machine learning*, pages 1–8. ACM, 2005.

[AN15] Michael A. Nielsen. *Neural networks and deep learning*. Determination Press, 2015.

[AR01] David Andre and Stuart J Russell. Programmable reinforcement learning agents. 2001.

[Aru10] E Arunan. Peter Debye. *Resonance: Journal of Science Education*, 15(12):1056–1059, 2010.

[AS97] Christopher G Atkeson and Juan Carlos Santamaria. A comparison of direct and model-based reinforcement learning. In *In International Conference on Robotics and Automation*. Citeseer, 1997.

[aut] Information of autoclave. http://www.aschome.com/index.php/en/. Accessed: 2015-02-03.

[B+95] Leemon Baird et al. Residual algorithms: Reinforcement learning with function approximation. In *ICML*, pages 30–37, 1995.

[B+06] Christopher M Bishop et al. *Pattern recognition and machine learning*, volume 1. springer New York, 2006.

[BADAC92] JGP Binner, IA Al-Dawery, C Aneziris, and TE Cross. Use of the inverse temperature profile in microwave processing of advanced ceramics. In *MRS Proceedings*, volume 269, page 357. Cambridge Univ Press, 1992.

[Bar89] Horace B Barlow. Unsupervised learning. *Neural computation*, 1(3):295–311, 1989.

[Bar94] Andrew G Barto. Reinforcement learning control. *Current Opinion in Neurobiology*, 4(6):888–893, 1994.

[Bar98] Andrew G Barto. *Reinforcement learning: An introduction*. MIT press, 1998.

[Bar01] Michael Barr. Pulse width modulation. *Embedded Systems Programming*, 14(10):103–104, 2001.

[Bat92] Roberto Battiti. First-and second-order methods for learning: between steepest descent and Newton's method. *Neural computation*, 4(2):141–166, 1992.

[BBBB95] Dimitri P Bertsekas, Dimitri P Bertsekas, Dimitri P Bertsekas, and Dimitri P Bertsekas. *Dynamic programming and optimal control*, volume 1. Athena Scientific Belmont, MA, 1995.

[BBG+03] SW Berg, D Bromberek, J Gagliano, AE Grelick, G Goeppner, A Nassiri, and T Smith. Rebuilding WR-340 and WR-284 waveguide switches to meet higher power at the advanced photon source. In *Particle Accelerator Conference, 2003. PAC 2003. Proceedings of the*, volume 2, pages 1416–1418. IEEE, 2003.

[BC06] Bruno Bouzy and Guillaume Chaslot. Monte-Carlo Go reinforcement learning experiments. In *Computational Intelligence and Games, 2006 IEEE Symposium on*, pages 187–194. IEEE, 2006.

[BDVW97] S Bradshaw, S Delport, and E Van Wyk. Qualitative measurement of heating uniformity in a multimode microwave cavity. *Journal of microwave power and electromagnetic energy*, 32(2):88–95, 1997.

[Bel57] Richard Bellman. A Markovian decision process. Technical report, DTIC Document, 1957.

[BG97] Zvi Boger and Hugo Guterman. Knowledge extraction from artificial neural network models. In *Sys-*

tems, Man, and Cybernetics, 1997. Computational Cybernetics and Simulation., 1997 IEEE International Conference on, volume 4, pages 3030–3035. IEEE, 1997.

[BH94] Martin Brown and Christopher John Harris. Neurofuzzy adaptive modelling and control. 1994.

[BJR13] George EP Box, Gwilym M Jenkins, and Gregory C Reinsel. *Time series analysis: forecasting and control.* John Wiley & Sons, 2013.

[BK93] Leemon C Baird and A Harry Klopf. Reinforcement learning with high-dimensional, continuous actions. *US Air Force Technical Report WL-TR-93-1147, Wright Laboratory, Wright-Patterson Air Force Base, OH*, 1993.

[BK04] Shalabh Bhatnagar and Shishir Kumar. A simultaneous perturbation stochastic approximation-based actor-critic algorithm for Markov decision processes. *Automatic Control, IEEE Transactions on*, 49(4):592–598, 2004.

[BM95] Justin Boyan and Andrew W Moore. Generalization in reinforcement learning: Safely approximating the value function. *Advances in neural information processing systems*, pages 369–376, 1995.

[BM03] Andrew G Barto and Sridhar Mahadevan. Recent advances in hierarchical reinforcement learning. *Discrete Event Dynamic Systems*, 13(4):341–379, 2003.

[Bot10] Léon Bottou. Large-scale machine learning with stochastic gradient descent. In *Proceedings of COMPSTAT'2010*, pages 177–186. Springer, 2010.

[Bow99] JR Bows. Variable frequency microwave heating of food. *Journal of Microwave Power and Electromagnetic Energy*, 34(4):227–238, 1999.

[Boy02] Justin A Boyan. Technical update: Least-squares temporal difference learning. *Machine Learning*, 49(2-3):233–246, 2002.

[BPJD99] John R Bows, Maria L Patrick, Ruth Janes, and David C Dibben. Microwave phase control heating. *International journal of food science & technology*, 34(4):295–304, 1999.

[Bri12] David R Brillinger. A generalized linear model with Gaussian regressor variables. In *Selected Works of David Brillinger*, pages 589–606. Springer, 2012.

[BRS01] Yu V Bykov, KI Rybakov, and VE Semenov. High-temperature microwave processing of materials. *Journal of Physics D: Applied Physics*, 34(13):R55, 2001.

[BSB81] Andrew G Barto, Richard S Sutton, and Peter S Brouwer. Associative search network: A reinforcement learning associative memory. *Biological cybernetics*, 40(3):201–211, 1981.

[BSF94] Yoshua Bengio, Patrice Simard, and Paolo Frasconi. Learning long-term dependencies with gradient descent is difficult. *Neural Networks, IEEE Transactions on*, 5(2):157–166, 1994.

[Bur98] Jeff B Burl. *Linear Optimal Control: H (2) and H (Infinity) Methods*. Addison-Wesley Longman Publishing Co., Inc., 1998.

[BV03] Hamid R Berenji and David Vengerov. A convergent actor-critic-based FRL algorithm with application to power management of wireless transmitters. *Fuzzy Systems, IEEE Transactions on*, 11(4):478–485, 2003.

[BV04] Stephen Boyd and Lieven Vandenberghe. *Convex optimization*. Cambridge university press, 2004.

[CA13] Eduardo F Camacho and Carlos Bordons Alba. *Model predictive control*. Springer, 2013.

[Cas12] EG Cason. The future of carbon fiber to 2017: global market forecasts. *Leatherhead, UK*, 2012.

[CBCB04] Eduardo F Camacho, Carlos Bordons, Eduardo F Camacho, and Carlos Bordons. *Model predictive control*, volume 2. Springer London, 2004.

[CC14] Wei-Der Chang and Chih-Yung Chen. PID controller design for MIMO processes using improved particle swarm optimization. *Circuits, Systems, and Signal Processing*, 33(5):1473–1490, 2014.

[CCHB89] Scott H Clearwater, Tze-Pin Cheng, Haym Hirsh, and Bruce G Buchanan. Incremental batch learning. In *Proceedings of the sixth international workshop on Machine learning*, pages 366–370. Morgan Kaufmann Publishers Inc., 1989.

[cfr] Price of CFRP per kilogram for the year 2014. http://www.reuters.com/article/2014/03/28/sgl-fibres-idUSL5N0MP2RP20140328. Accessed: 2015-02-09.

[Che03] Wen-Hua Chen. Closed-form nonlinear mpc for multivariable nonlinear systems with different relative degree. In *American Control Conference, 2003. Proceedings of the 2003*, volume 6, pages 4887–4892. IEEE, 2003.

[Chr05] Christos Christopoulos. The transmission-line modeling (TLM) method in electromagnetics. *Synthesis Lectures on Computational Electromagnetics*, 1(1):1–132, 2005.

[CK92] Fu-Chuang Chen and Hassan K Khalil. Adaptive control of nonlinear systems using neural networks. *International Journal of Control*, 55(6):1299–1317, 1992.

[CMA94] Jerome T Connor, R Douglas Martin, and Les E Atlas. Recurrent neural networks and robust time series prediction. *Neural Networks, IEEE Transactions on*, 5(2):240–254, 1994.

[CMT87] David W Clarke, C Mohtadi, and PS Tuffs. Generalized predictive controlpart i. the basic algorithm. *Automatica*, 23(2):137–148, 1987.

[Cog92] Frederic Neil Cogswell. *Thermoplastic aromatic polymer composites: a study of the structure, processing and properties of carbon fibre reinforced polyetheretherketone and related materials*. Elsevier, 1992.

[Cor01] JM Corum. Basic properties of of reference crossply carbon-fiber composite. Technical report, Oak Ridge National Lab., TN (US), 2001.

[CRB08] S Curet, O Rouaud, and L Boillereaux. Microwave tempering and heating in a single-mode cavity: Numerical and experimental investigations. *Chemical Engineering and Processing: Process Intensification*, 47(9):1656–1665, 2008.

[Cro81] Lawrence Crone. Second order adjoint matrix equations. *Linear Algebra and its Applications*, 39:61–71, 1981.

[CS93] Hown-Wen Chen and Von-Wun Soo. Design of adaptive and incremental feed-forward neural networks. In *Neural Networks, 1993., IEEE International Conference on*, pages 479–484. IEEE, 1993.

[CS02] Tien-Chi Chen and Tsong-Terng Sheu. Model reference neural network controller for induction motor speed control. *Energy Conversion, IEEE Transactions on*, 17(2):157–163, 2002.

[CU93] A Cochocki and Rolf Unbehauen. *Neural networks for optimization and signal processing*. John Wiley & Sons, Inc., 1993.

[Cut83] Charles Ray Cutler. Dynamic matrix control: An optimal multivariable control algorithm with constraints. 1983.

[CYC02] Shi-Yuan Chen, Fang-Ming Yu, and Hung-Yuan Chung. Decoupled fuzzy controller design with single-input fuzzy logic. *Fuzzy Sets and Systems*, 129(3):335–342, 2002.

[CZ05] LA Campanone and NE Zaritzky. Mathematical analysis of microwave heating process. *Journal of Food Engineering*, 69(3):359–368, 2005.

[DAF90] AAP De Alwis and PJ Fryer. The use of direct resistance heating in the food industry. *Journal of Food Engineering*, 11(1):3–27, 1990.

[DAJ02] Kalyanmoy Deb, Ashish Anand, and Dhiraj Joshi. A computationally efficient evolutionary algorithm for real-parameter optimization. *Evolutionary computation*, 10(4):371–395, 2002.

[Dat01] Ashim K Datta. *Handbook of microwave technology for food application*. CRC Press, 2001.

[DC03] Danielle Dougherty and Doug Cooper. A practical multiple model adaptive strategy for multivariable model predictive control. *Control Engineering Practice*, 11(6):649–664, 2003.

[dFNGD00] João FG de Freitas, Mahesan Niranjan, Andrew H. Gee, and Arnaud Doucet. Sequential Monte Carlo methods to train neural network models. *Neural computation*, 12(4):955–993, 2000.

[Dib95] David Dibben. *Numerical and experimental modelling of microwave applicators*. PhD thesis, University of Cambridge, 1995.

[Die98] Thomas G Dietterich. The MAXQ method for hierarchical reinforcement learning. In *ICML*, pages 118–126. Citeseer, 1998.

[Die99] Thomas G Dietterich. Hierarchical reinforcement learning with the MAXQ value function decomposition. *arXiv preprint cs/9905014*, 1999.

[DKS+95] James Dougherty, Ron Kohavi, Mehran Sahami, et al. Supervised and unsupervised discretization of continuous features. In *ICML*, pages 194–202, 1995.

[DLT12] Gouri Dhatt, Emmanuel Lefrançois, and Gilbert Touzot. *Finite element method*. John Wiley & Sons, 2012.

[DM07] Ernest O Doebelin and Dhanesh N Manik. Measurement systems: Application and design. 2007.

[Dor95] Richard C Dorf. *Modern control systems*. Addison-Wesley Longman Publishing Co., Inc., 1995.

[DPP02] Duško D Dinčov, Kevin A Parrott, and Koulis A Pericleous. Coupled 3-D finite difference time domain and finite volume methods for solving microwave heating in porous media. In *Computational Science - ICCS 2002*, pages 813–822. Springer, 2002.

[DSKK02] Kenji Doya, Kazuyuki Samejima, Ken-ichi Katagiri, and Mitsuo Kawato. Multiple model-based reinforcement learning. *Neural computation*, 14(6):1347–1369, 2002.

[DWA+05] K. Darcovich, P.S. Whitfield, G. Amow, K. Shinagawa, and R.Y. Miyahara. A microstructure based numerical simulation of microwave sintering of specialized SOFC materials. *Journal of the European Ceramic Society*, 25(12):2235 – 2240, 2005.

[EDH05] D Edouard, Pascal Dufour, and Hassan Hammouri. Observer based multivariable control of a catalytic reverse flow reactor: comparison between LQR and MPC approaches. *Computers & chemical engineering*, 29(4):851–865, 2005.

[EEE97] Delores M Etter, Dolores M Etter, and Delores M Etter. *Engineering problem solving with MATLAB*. Prentice Hall New Jersey, 1997.

[Elm90] Jeffrey L Elman. Finding structure in time. *Cognitive science*, 14(2):179–211, 1990.

[EUDC07] Stefan Elfwing, Eiji Uchibe, Kenji Doya, and Henrik I Christensen. Evolutionary development of hierarchical learning structures. *Evolutionary Computation, IEEE Transactions on*, 11(2):249–264, 2007.

[FA02] Rolf Findeisen and Frank Allgöwer. An introduction to nonlinear model predictive control. In *21st Benelux Meeting on Systems and Control*, volume 11, 2002.

[FBLI+95] Arne W Fliflet, Ralph W Bruce, David Lewis III, R Rayne, and B Bender. Design and initial operation of a 6 kW 2.45 GHz single-mode microwave cavity furnace. Technical report, DTIC Document, 1995.

[FDH01] Ray J Frank, Neil Davey, and Stephen P Hunt. Time series prediction and neural networks. *Journal of Intelligent and Robotic Systems*, 31(1-3):91–103, 2001.

[Feh09] L Feher. *Energy Efficient Microwave Systems*. Springer, 2009.

[FKÖS11] Semuel Franko, İlker M Koç, Can Özsoy, and Nurşen Sari. MPC and LQR type controller design and comparison for an unmanned helicopter. In *Proceedings of the 2011 Summer Computer Simulation Conference*, pages 138–144. Society for Modeling & Simulation International, 2011.

[fli] Introduction of FLIR thermovision toolkit. http://www.flir.com/thermography/americas/us/content/?id=18276. Accessed: 2015-02-03.

[FR93] M.O. Fakhouri and H.S. Ramaswamy. Temperature uniformity of microwave heated foods as influenced by product type and composition. *Food Research International*, 26(2):89 – 95, 1993.

[Fre99] Robert M French. Catastrophic forgetting in connectionist networks. *Trends in cognitive sciences*, 3(4):128–135, 1999.

[FT88] Glenville CE Fowler and David J Tozer. Inter-bus system, March 1 1988. US Patent 4,728,754.

[GA11] Pascal Gahinet and Pierre Apkarian. Decentralized and fixed-structure H-infinity control in MATLAB. In *Decision and Control and European Control Conference (CDC-ECC), 2011 50th IEEE Conference on*, pages 8205–8210. IEEE, 2011.

[GBLB12] Ivo Grondman, Lucian Busoniu, Gabriel AD Lopes, and Robert Babuska. A survey of actor-critic reinforcement learning: Standard and natural policy gradients. *Systems, Man, and Cybernetics, Part C: Applications and Reviews, IEEE Transactions on*, 42(6):1291–1307, 2012.

[GBTH06] Nicolò Giorgetti, Alberto Bemporad, H Eric Tseng, and Davor Hrovat. Hybrid model predictive control application towards optimal semi-active suspension. *International Journal of Control*, 79(05):521–533, 2006.

[GC99] David Jeffrey Griffiths and Reed College. *Introduction to electrodynamics*, volume 3. Prentice Hall Upper Saddle River, NJ, 1999.

[GHLZ10] SS Ge, Chang C Hang, Tong H Lee, and Tao Zhang. *Stable adaptive neural network control*. Springer Publishing Company, Incorporated, 2010.

[Gil11] AS Gilmour. *Klystrons, traveling wave tubes, magnetrons, crossed-field amplifiers, and gyrotrons*. Artech House, 2011.

[Glo93] Pierre Yves Glorennec. Adaptive fuzzy control. In *Fuzzy Logic*, pages 541–551. Springer, 1993.

[GM82] Carlos E Garcia and Manfred Morari. Internal model control. a unifying review and some new results. *Industrial & Engineering Chemistry Process Design and Development*, 21(2):308–323, 1982.

[GM86] Carlos E Garcia and AM Morshedi. Quadratic programming solution of dynamic matrix control (QDMC). *Chemical Engineering Communications*, 46(1-3):73–87, 1986.

[Gmb14] Voetsch Industrietechnik GmbH. Energy efficient microwave curing of carbon fibre reinforced plastics (CFRP) first serial sandwich part for aerospace cured with microwave system HEPHAISTOS VHM. *Technical Press Release of Voetsch Industrietechnik GmbH*, 2014.

[GPM89] Carlos E Garcia, David M Prett, and Manfred Morari. Model predictive control: Theory and practice - a survey. *Automatica*, 25(3):335–348, 1989.

[GRD07] SSR Geedipalli, V Rakesh, and AK Datta. Modeling the heating uniformity contributed by a rotating turntable in microwave ovens. *Journal of Food Engineering*, 82(3):359–368, 2007.

[Gru03] Klaus-Dieter Gruner. Principles of non-contact temperature measurement. 2003.

[GWZ99] Chris Gaskett, David Wettergreen, and Alexander Zelinsky. Q-learning in continuous state and action spaces. In *Advanced Topics in Artificial Intelligence*, pages 417–428. Springer, 1999.

[GY07] Sundaram Gunasekaran and Huai-Wen Yang. Effect of experimental parameters on temperature distribution during continuous and pulsed microwave heating. *Journal of Food Engineering*, 78(4):1452 – 1456, 2007.

[Hai01] Richard E Haimbaugh. *Practical induction heat treating*. Asm International, 2001.

[Hay98] Simon Haykin. *Neural networks: a comprehensive foundation*. Prentice Hall PTR, 1998.

[HB00] MN Howell and Matt C Best. On-line PID tuning for engine idle-speed control using continuous action reinforcement learning automata. *Control Engineering Practice*, 8(2):147–154, 2000.

[HB01] William Hart Hayt and John A Buck. *Engineering electromagnetics*, volume 7. McGraw-Hill New York, 2001.

[HBC95] NA Hassine, JGP Binner, and TE Cross. Synthesis of refractory metal carbide powders via microwave carbothermal reduction. *International Journal of Refractory Metals and Hard Materials*, 13(6):353–358, 1995.

[HBS11] Robert Haber, Ruth Bars, and Ulrich Schmitz. Predictive functional control. *Predictive Control in Process Engineering: From the Basics to the Applications*, pages 437–465, 2011.

[HDB+96] Martin T Hagan, Howard B Demuth, Mark H Beale, et al. *Neural network design*, volume 1. Pws Boston, 1996.

[HJ93] James M. Hill and Michael J. Jennings. Formulation of model equations for heating by microwave radiation. *Applied Mathematical Modelling*, 17(7):369 – 379, 1993.

[HJ07] Pingan He and Sarangapani Jagannathan. Reinforcement learning neural-network-based controller for nonlinear discrete-time systems with input constraints. *Systems, Man, and Cybernetics, Part B: Cybernetics, IEEE Transactions on*, 37(2):425–436, 2007.

[HJK95] Christopher R Houck, Jeffery A Joines, and Michael G Kay. A genetic algorithm for function optimization: A matlab implementation. *NCSU-IE TR*, 95(09), 1995.

[HM96] James M Hill and Timothy R Marchant. Modelling microwave heating. *Applied Mathematical Modelling*, 20(1):3 – 15, 1996.

[Hop82] John J Hopfield. Neural networks and physical systems with emergent collective computational abilities. *Proceedings of the national academy of sciences*, 79(8):2554–2558, 1982.

[Hor91] Kurt Hornik. Approximation capabilities of multilayer feedforward networks. *Neural networks*, 4(2):251–257, 1991.

[HPE97] Y Halevi, Zalman J Palmor, and T Efrati. Automatic tuning of decentralized PID controllers for MIMO processes. *Journal of Process Control*, 7(2):119–128, 1997.

[HS92] Robert E Hinchee and Lawrence A Smith. *In situ thermal technologies for site remediation*. CRC Press, 1992.

[HS07] Lihan Huang and Joseph Sites. Automatic control of a microwave heating process for in-package pasteurization of beef frankfurters. *Journal of food engineering*, 80(1):226–233, 2007.

[HSW89] Kurt Hornik, Maxwell Stinchcombe, and Halbert White. Multilayer feedforward networks are universal approximators. *Neural networks*, 2(5):359–366, 1989.

[IKI02] Jun Izawa, Toshiyuki Kondo, and Koji Ito. Biological robot arm motion through reinforcement learning. In *SICE 2002. Proceedings of the 41st SICE Annual Conference*, volume 1, pages 413–418. IEEE, 2002.

[IMMC08] Engin Ipek, Onur Mutlu, José F Martínez, and Rich Caruana. Self-optimizing memory controllers: A reinforcement learning approach. In *Computer Architecture, 2008. ISCA'08. 35th International Symposium on*, pages 39–50. IEEE, 2008.

[Ind13] Despatch Industries. Complete composite curing systems: out-of-autoclave solutions that meet the most demanding aerospace requirements. Technical report, Despatch Industries, 2013.

[IYY02] Shin Ishii, Wako Yoshida, and Junichiro Yoshimoto. Control of exploitation–exploration metaparameter in reinforcement learning. *Neural networks*, 15(4):665–687, 2002.

[JF05] Jürgen Jasperneite and Joachim Feld. PROFINET: an integration platform for heterogeneous industrial communication systems. In *Emerging Technologies and Factory Automation, 2005. ETFA 2005. 10th IEEE Conference on*, volume 1, pages 8–pp. IEEE, 2005.

[JHB+95] Anatoli Juditsky, Håkan Hjalmarsson, Albert Benveniste, Bernard Delyon, Lennart Ljung, Jonas Sjöberg, and Qinghua Zhang. Nonlinear black-box models in system identification: Mathematical foundations. *Automatica*, 31(12):1725–1750, 1995.

[JLL00] Chia-Feng Juang, Jiann-Yow Lin, and Chin-Teng Lin. Genetic reinforcement learning through symbiotic evolution for fuzzy controller design. *Systems, Man, and Cybernetics, Part B: Cybernetics, IEEE Transactions on*, 30(2):290–302, 2000.

[JMM96] Anil K Jain, Jianchang Mao, and KM Mohiuddin. Artificial neural networks: A tutorial. *Computer*, 29(3):31–44, 1996.

[Jon98] Donnie Jones. Navy electricity and electronics training series - introduction to electronic emission, tubes, and power supplies. *Naval Education and Training Professional Development and Technnology Center*, 1998.

[JsK07] Seul Jung and Sung su Kim. Hardware implementation of a real-time neural network controller with a DSP and an FPGA for nonlinear systems. *Industrial Electronics, IEEE Transactions on*, 54(1):265–271, 2007.

[Kat92] Joel D Katz. Microwave sintering of ceramics. *Annual Review of Materials Science*, 22(1):153–170, 1992.

[KBM96] Mayuresh V Kothare, Venkataramanan Balakrishnan, and Manfred Morari. Robust constrained model predictive control using linear matrix inequalities. *Automatica*, 32(10):1361–1379, 1996.

[KHL+12] Said G Khan, Guido Herrmann, Frank L Lewis, Tony Pipe, and Chris Melhuish. Reinforcement learning and optimal adaptive control: An overview and implementation examples. *Annual Reviews in Control*, 36(1):42–59, 2012.

[KHYM13] Mitsuhiro Kubota, Takuya Hanada, Satoshi Yabe, and Hitoki Matsuda. Regeneration characteristics of desiccant rotor with microwave and hot-air heating. *Applied Thermal Engineering*, 50(2):1576–1581, 2013.

[KL93] Karl S Kunz and Raymond J Luebbers. *The finite difference time domain method for electromagnetics*. CRC press, 1993.

[KLM96] Leslie Pack Kaelbling, Michael L Littman, and Andrew W Moore. Reinforcement learning: A survey. *arXiv preprint cs/9605103*, 1996.

[KMS01] Magdalini K Krokida, Zacharias B Maroulis, and George D Saravacos. The effect of the method of drying on the color of dehydrated products. *International journal of food science & technology*, 36(1):53–59, 2001.

[KO92] Marzuki Khalid and Sigeru Omatu. A neural network controller for a temperature control system. *Control Systems, IEEE*, 12(3):58–64, 1992.

[Kri92] Gregory A Kriegsmann. Thermal runaway in microwave heated ceramics: A one-dimensional model. *Journal of Applied Physics*, 71(4):1960–1966, 1992.

[KRS08] Kai Knoerzer, Marc Regier, and Helmar Schubert. A computational model for calculating temperature distributions in microwave food applications. *Innovative Food Science & Emerging Technologies*, 9(3):374–384, 2008.

[KV86] Panqanamala Ramana Kumar and Pravin Varaiya. *Stochastic systems: Estimation, identification, and adap-*

tive control. Prentice Hall Englewood Cliffs, NJ, 1986.

[KV92] Nicolaos Karayiannis and Anastasios N Venetsanopoulos. *Artificial neural networks: learning algorithms, performance evaluation, and applications*, volume 209. Springer, 1992.

[KY97] Harold J Kushner and G George Yin. *Stochastic approximation algorithms and applications.* Springer, 1997.

[lab] Official website of LabVIEW. http://www.ni.com/labview/. Accessed: 2015-02-03.

[lar] Report of world's largest microwave heating device. http://www.netcomposites.com/news/worlds-largest-industrial-composite-microwave-system/5366. Accessed: 2015-02-06.

[LBC+95] B Levush, A Birman, Y Carmel, M Rosen, and D Abe. Multi-frequency microwave sintering of ceramics. In *Plasma Science, 1995. IEEE Conference Record-Abstracts., 1995 IEEE International Conference on*, page 213. IEEE, 1995.

[Lee90] Chuen-Chien Lee. Fuzzy logic in control systems: fuzzy logic controller. ii. *Systems, Man and Cybernetics, IEEE Transactions on*, 20(2):419–435, 1990.

[Li08] Ta-Hsin Li. On exponentially weighted recursive least squares for estimating time-varying parameters and its application to computer workload forecasting. *Journal of Statistical Theory and Practice*, 2(3):339–354, 2008.

[Lin93] Long-Ji Lin. Reinforcement learning for robots using neural networks. Technical report, DTIC Document, 1993.

[Lin03] Chuan-Kai Lin. A reinforcement learning adaptive fuzzy controller for robots. *Fuzzy sets and systems*, 137(3):339–352, 2003.

[LJ00] Chin-Teng Lin and Chong-Ping Jou. GA-based fuzzy reinforcement learning for control of a magnetic bearing system. *Systems, Man, and Cybernetics, Part B: Cybernetics, IEEE Transactions on*, 30(2):276–289, 2000.

[Lju97] Lennart Ljung. Nonlinear black box models in system identification. In *Proceedings of the 1997 IFAC International Symposum on Advanced Control of Chemical Processes*, pages 1–13, 1997.

[Lju98] Lennart Ljung. *System identification*. Springer, 1998.

[LLG14] Yingguang Li, Nanya Li, and James Gao. Tooling design and microwave curing technologies for the manufacturing of fiber-reinforced polymer composites in aerospace applications. *The International Journal of Advanced Manufacturing Technology*, 70(1-4):591–606, 2014.

[LLHG14] Nanya Li, Yingguang Li, Xiaozhong Hao, and James Gao. A comparative experiment for the analysis of microwave and thermal process induced strains of carbon fiber/bismaleimide composite materials. *Composites Science and Technology*, 2014.

[LM03] Nicholas E Leadbeater and Maria Marco. Rapid and amenable Suzuki coupling reaction in water using microwave and conventional heating. *The Journal of organic chemistry*, 68(3):888–892, 2003.

[LMAC02] D Limón Marruedo, T Alamo, and EF Camacho. Stability analysis of systems with bounded additive uncertainties based on invariant sets: Stability and feasibility of MPC. In *American Control Conference, 2002. Proceedings of the 2002*, volume 1, pages 364–369. IEEE, 2002.

[LN95] Asriel U Levin and Kumpati S Narendra. Control of nonlinear dynamical systems using neural networks. ii. observability, identification, and control.

IEEE transactions on neural networks/a publication of the IEEE Neural Networks Council, 7(1):30–42, 1995.

[LNS04] Roland Wynne Lewis, Perumal Nithiarasu, and Kankanhalli Seetharamu. *Fundamentals of the finite element method for heat and fluid flow*. John Wiley & Sons, 2004.

[LP02] Mircea Lazar and Octavian Pastravanu. A neural predictive controller for non-linear systems. *Mathematics and Computers in Simulation*, 60(3):315–324, 2002.

[LR95] Peter Lancaster and Leiba Rodman. *Algebraic riccati equations*. Oxford University Press, 1995.

[LS98] John Loch and Satinder P Singh. Using eligibility traces to find the best memoryless policy in partially observable Markov decision processes. In *ICML*, pages 323–331, 1998.

[LV09] Frank L Lewis and Draguna Vrabie. Reinforcement learning and adaptive dynamic programming for feedback control. *Circuits and Systems Magazine, IEEE*, 9(3):32–50, 2009.

[LY97] JH a Lee and Zhenghong Yu. Worst-case formulations of model predictive control for systems with bounded parameters. *Automatica*, 33(5):763–781, 1997.

[Mad04] Robert P Madding. Infrared window transmittance temperature dependence. *InfraMation 2004 Proceedings*, 2004.

[mat] Official website of MATLAB. http://www.mathworks.com/products/matlab/. Accessed: 2015-02-03.

[MC89] Michael McCloskey and Neal J Cohen. Catastrophic interference in connectionist networks: The sequential learning problem. *Psychology of learning and motivation*, 24:109–165, 1989.

[Meh09] Mehrdad Mehdizadeh. *Microwave/RF applicators and probes for material heating, sensing, and plasma generation: a design guide.* William Andrew, 2009.

[Mer98] Roger J Meredith. *Engineers' handbook of industrial microwave heating.* Number 25. IET, 1998.

[Met96] AC Metaxas. *Foundations of electroheat: A unified approach.* John Wiley & Sons Inc, 1996.

[MHL99] Manfred Morari and Jay H Lee. Model predictive control: past, present and future. *Computers & Chemical Engineering*, 23(4):667–682, 1999.

[MJC04] Luiz Meyer, Shesha Jayaram, and Edward A Cherney. Thermal conductivity of filled silicone rubber and its relationship to erosion resistance in the inclined plane test. *Dielectrics and Electrical Insulation, IEEE Transactions on*, 11(4):620–630, 2004.

[MKJ02] MH Moradi, MR Katebi, and MA Johnson. The MIMO predictive PID controller design. *Asian Journal of Control*, 4(4):452–463, 2002.

[ML80] P Martin Larsen. Industrial applications of fuzzy logic control. *International Journal of Man-Machine Studies*, 12(1):3–10, 1980.

[MM83] AC Metaxas, , and Roger J Meredith. *Industrial microwave heating.* Number 4. IET, 1983.

[Mor85] Manfred Morari. Robust stability of systems with integral control. *Automatic Control, IEEE Transactions on*, 30(6):574–577, 1985.

[MRRS00] David Q Mayne, James B Rawlings, Christopher V Rao, and Pierre OM Scokaert. Constrained model predictive control: Stability and optimality. *Automatica*, 36(6):789–814, 2000.

[MSH98] Xiao-Jun Ma, Zeng-Qi Sun, and Yan-Yan He. Analysis and design of fuzzy controller and fuzzy observer. *Fuzzy Systems, IEEE Transactions on*, 6(1):41–51, 1998.

[MTA+06] John JL Morton, Alexei M Tyryshkin, Arzhang Ardavan, Simon C Benjamin, Kyriakos Porfyrakis, SA Lyon, and G Andrew D Briggs. Bang–bang control of fullerene qubits using ultrafast phase gates. *Nature Physics*, 2(1):40–43, 2006.

[Mur00] Nial Murphy. Watchdog timers. *Embedded Systems Programming*, page 112, 2000.

[NELL08] Christopher Niedzwiedz, Itamar Elhanany, Zhenzhen Liu, and Scott Livingston. A consolidated actor-critic model with function approximation for high-dimensional POMDPs. In *AAAI 2008 workshop for Advancement in POMDP Solvers (part of the AAAI 2008 Conference), Chicago*, 2008.

[NP97] Vesna Nevistić and James A Primbs. Finite receding horizon linear quadratic control: A unifying theory for stability and performance analysis. 1997.

[NSV98] Lester SH Ngia, Jonas Sjoberg, and Mats Viberg. Adaptive neural nets filter using a recursive Levenberg-Marquardt search direction. In *Signals, Systems & Computers, 1998. Conference Record of the Thirty-Second Asilomar Conference on*, volume 1, pages 697–701. IEEE, 1998.

[ODF+03] John P O'Doherty, Peter Dayan, Karl Friston, Hugo Critchley, and Raymond J Dolan. Temporal difference models and reward-related learning in the human brain. *Neuron*, 38(2):329–337, 2003.

[oED07] U.S. Department of Energy (DOE). *Improving process heating system performance: A sourcebook for industry.* DOE, 2007.

[OM10] Morteza Oghbaei and Omid Mirzaee. Microwave versus conventional sintering: A review of fundamentals, advantages and applications. *Journal of Alloys and Compounds*, 494(1):175–189, 2010.

[OOH95] M Ohshima, H Ohno, and I Hashimoto. Model predictive control: Experiences in the university-industry joint projects and statistics on MPC applications in Japan. In *International workshop on predictive and receding horizon control*, pages 1–16, 1995.

[PCB+01] Evan Pert, Yuval Carmel, Amikam Birnboim, Tayo Olorunyolemi, David Gershon, Jeff Calame, Isabel K Lloyd, and Otto C Wilson. Temperature measurements during microwave processing: The significance of thermocouple effects. *Journal of the American Ceramic Society*, 84(9):1981–1986, 2001.

[PGMCCCSH04] Pedro Plaza-González, Juan Monzó-Cabrera, José M Catalá-Civera, and David Sánchez-Hernández. New approach for the prediction of the electric field distribution in multimode microwave-heating applicators with mode stirrers. *Magnetics, IEEE Transactions on*, 40(3):1672–1678, 2004.

[PKM13] Shashi Kant Pandey, Nand Kishor, and Soumya Ranjan Mohanty. A novel decentralized robust MIMO-PID load frequency controller via iterative lmi approach in deregulated environment. In *Engineering and Systems (SCES), 2013 Students Conference on*, pages 1–6. IEEE, 2013.

[PL05] Charles Pandana and KJ Ray Liu. Near-optimal reinforcement learning framework for energy-aware sensor communications. *Selected Areas in Communications, IEEE Journal on*, 23(4):788–797, 2005.

[Pol03] DSG Pollock. Recursive estimation in econometrics. *Computational statistics & data analysis*, 44(1):37–75, 2003.

[Poz09] David M Pozar. *Microwave engineering*. John Wiley & Sons, 2009.

[PPW+02] RD Patton, CU Pittman, L Wang, JR Hill, and A Day. Ablation, mechanical and thermal conductivity properties of vapor grown carbon fiber/phenolic matrix composites. *Composites Part A: Applied Science and Manufacturing*, 33(2):243–251, 2002.

[PR99] Vassilis A Papavassiliou and Stuart Russell. Convergence of reinforcement learning with general function approximators. In *IJCAI*, pages 748–757, 1999.

[PS91] AH Pincombe and NF Smyth. Microwave heating of materials with low conductivity. *Proceedings of the Royal Society of London. Series A: Mathematical and Physical Sciences*, 433(1889):479–498, 1991.

[PS06] Jan Peters and Stefan Schaal. Policy gradient methods for robotics. In *Intelligent Robots and Systems, 2006 IEEE/RSJ International Conference on*, pages 2219–2225. IEEE, 2006.

[PS08a] Jan Peters and Stefan Schaal. Natural actor-critic. *Neurocomputing*, 71(7):1180–1190, 2008.

[PS08b] Jan Peters and Stefan Schaal. Reinforcement learning of motor skills with policy gradients. *Neural networks*, 21(4):682–697, 2008.

[PSD01] Doina Precup, Richard S Sutton, and Sanjoy Dasgupta. Off-policy temporal-difference learning with function approximation. In *ICML*, pages 417–424, 2001.

[PSRJG00] Stephen Piche, Bijan Sayyar-Rodsari, Doug Johnson, and Mark Gerules. Nonlinear model predictive control using neural networks. *Control Systems, IEEE*, 20(3):53–62, 2000.

[PVS05] Jan Peters, Sethu Vijayakumar, and Stefan Schaal. Natural actor-critic. In *Machine Learning: ECML 2005*, pages 280–291. Springer, 2005.

[PW96] Jing Peng and Ronald J Williams. Incremental multi-step Q-learning. *Machine Learning*, 22(1-3):283–290, 1996.

[PW08] Yunpeng Pan and Jun Wang. Nonlinear model predictive control using a recurrent neural network. In *Neural Networks, 2008. IJCNN 2008.(IEEE World*

Congress on Computational Intelligence). IEEE International Joint Conference on, pages 2296–2301. IEEE, 2008.

[QB03] S Joe Qin and Thomas A Badgwell. A survey of industrial model predictive control technology. *Control engineering practice*, 11(7):733–764, 2003.

[Qui01] Adam Quilter. Composites in aerospace applications. *IHS White Paper*, 444:1–3, 2001.

[RCVI99] VG Ryckaert, JE Claes, and JF Van Impe. Model-based temperature control in ovens. *Journal of food engineering*, 39(1):47–58, 1999.

[Rib04] Maria Isabel Ribeiro. Kalman and extended Kalman filters: Concept, derivation and properties. *Institute for Systems and Robotics*, page 43, 2004.

[Ric05] Arthur Richards. Robust model predictive control for time-varying systems. In *Decision and Control, 2005 and 2005 European Control Conference. CDC-ECC'05. 44th IEEE Conference on*, pages 3747–3752. IEEE, 2005.

[rip] Information of standard-ripple power supplies. http://cobermuegge.com/details.asp?id=132. Accessed: 2015-03-06.

[RLCB02] Valery Rudnev, Don Loveless, Raymond L Cook, and Micah Black. *Handbook of induction heating*. CRC Press, 2002.

[RLRJGA06] II Ruiz-López, GC Rodríguez-Jimenes, and MA García-Alvarado. Robust MIMO PID controllers tuning based on complex/real ratio of the characteristic matrix eigenvalues. *Chemical Engineering Science*, 61(13):4332–4340, 2006.

[RMS96] C García Reimbert, AA Minzoni, and NF Smyth. Effect of radiation losses on hotspot formation and propagation in microwave heating. *IMA journal of applied mathematics*, 57(2):165–179, 1996.

[RMTV85] G Roussy, A Mercier, JM Theibaut, and J-P Vaubourg. Temperature runaway of microwave heated materials: study and control. *Journal of microwave power*, 20(1):47–51, 1985.

[Rob95] Anthony Robins. Catastrophic forgetting, rehearsal and pseudorehearsal. *Connection Science*, 7(2):123–146, 1995.

[Roj96] Raúl Rojas. *Neural networks: a systematic introduction*. Springer Science & Business Media, 1996.

[RP98] G Roussy and JA Pearce. Foundations and industrial applications of microwaves and radio frequency fields. physical and chemical processes. In *Optimization of Electrical and Electronic Equipments, 1998. OPTIM'98. Proceedings of the 6th International Conference on*, volume 1, pages 115–116. IEEE, 1998.

[RP06] Edgar Rapoport and Yulia Pleshivtseva. *Optimal control of induction heating processes*. CRC Press, 2006.

[RRK+92] Dennis W. Ruck, Steven K. Rogers, Matthew Kabrisky, Peter S. Maybeck, and Mark E. Oxley. Comparative analysis of backpropagation and the extended Kalman filter for training multilayer perceptrons. *IEEE Transactions on Pattern Analysis and Machine Intelligence*, 14(6):686–691, 1992.

[RRTP78] Jacques Richalet, A Rault, JL Testud, and J Papon. Model predictive heuristic control: Applications to industrial processes. *Automatica*, 14(5):413–428, 1978.

[RS07] Rubén Ruiz and Thomas Stützle. A simple and effective iterated greedy algorithm for the permutation flowshop scheduling problem. *European Journal of Operational Research*, 177(3):2033–2049, 2007.

[Rug96] Wilson J Rugh. *Linear system theory*, volume 2. prentice hall Upper Saddle River, NJ, 1996.

[RZK91] GVS Raju, JUN Zhou, and Roger A Kisner. Hierarchical fuzzy control. *International journal of control*, 54(5):1201–1216, 1991.

[SAB98] José Saa, Antonio A Alonso, and Julio R Banga. Optimal control of microwave heating using mathematical models of medium complexity. *Automatic Control of Food and Biological Processes (AcoFop IV)*, 1998.

[SBA00] Isidro Sanchez, Julio R Banga, and Antonio A Alonso. Temperature control in microwave combination ovens. *Journal of Food Engineering*, 46(1):21–29, 2000.

[SBW92] Richard S Sutton, Andrew G Barto, and Ronald J Williams. Reinforcement learning is direct adaptive optimal control. *Control Systems, IEEE*, 12(2):19–22, 1992.

[SC94] James C Spall and John A Cristion. Nonlinear adaptive control using neural networks: Estimation with a smoothed form of simultaneous perturbation gradient approximation. In *American Control Conference, 1994*, volume 3, pages 2560–2564. IEEE, 1994.

[SC97] James C Spall and John A Cristion. A neural network controller for systems with unmodeled dynamics with applications to wastewater treatment. *Systems, Man, and Cybernetics, Part B: Cybernetics, IEEE Transactions on*, 27(3):369–375, 1997.

[SC98] James C Spall and John A Cristion. Model-free control of nonlinear stochastic systems with discrete-time measurements. *Automatic Control, IEEE Transactions on*, 43(9):1198–1210, 1998.

[Sch90] Jürgen Schmidhuber. An on-line algorithm for dynamic reinforcement learning and planning in reactive environments. In *Neural Networks, 1990., 1990 IJCNN International Joint Conference on*, pages 253–258. IEEE, 1990.

[Sch98] Alexander Schrijver. *Theory of linear and integer programming*. John Wiley & Sons, 1998.

[SCV+10] T Santos, LC Costa, M Valente, J Monteiro, and J Sousa. 3D electromagnetic field simulation in microwave ovens: A tool to control thermal runaway. In *COMSOL Conference*, 2010.

[Si04] Jennie Si. *Handbook of learning and approximate dynamic programming*, volume 2. John Wiley & Sons, 2004.

[Sie01] Robert Siegel. *Thermal radiation heat transfer*, volume 1. CRC press, 2001.

[Sil12] Shin-Etsu Silicone. Characteristic properties of silicone rubber compounds. 2012.

[SK00] William D Smart and Leslie Pack Kaelbling. Practical reinforcement learning in continuous spaces. In *ICML*, pages 903–910. Citeseer, 2000.

[SM96] Kenneth O Stanley and Risto Miikkulainen. Efficient reinforcement learning through evolving neural network topologies. *Network (Phenotype)*, 1(2):3, 1996.

[SM04] Kenneth O Stanley and Risto Miikkulainen. Competitive coevolution through evolutionary complexification. *J. Artif. Intell. Res.(JAIR)*, 21:63–100, 2004.

[SMDH13] Ilya Sutskever, James Martens, George Dahl, and Geoffrey Hinton. On the importance of initialization and momentum in deep learning. In *Proceedings of the 30th International Conference on Machine Learning (ICML-13)*, pages 1139–1147, 2013.

[SMSM99] Richard S Sutton, David A McAllester, Satinder P Singh, and Yishay Mansour. Policy gradient methods for reinforcement learning with function approximation. In *NIPS*, volume 99, pages 1057–1063. Citeseer, 1999.

[Smy90] NF Smyth. Microwave heating of bodies with temperature dependent properties. *Wave Motion*, 12(2):171–186, 1990.

[SP07] Sigurd Skogestad and Ian Postlethwaite. *Multivariable feedback control: analysis and design*, volume 2. Wiley New York, 2007.

[Spa87] James C Spall. A stochastic approximation technique for generating maximum likelihood parameter estimates. In *American Control Conference, 1987*, pages 1161–1167. IEEE, 1987.

[Spa97] James C Spall. A one-measurement form of simultaneous perturbation stochastic approximation. *Automatica*, 33(1):109–112, 1997.

[Spa98] James C Spall. Implementation of the simultaneous perturbation algorithm for stochastic optimization. *Aerospace and Electronic Systems, IEEE Transactions on*, 34(3):817–823, 1998.

[SS96] Satinder P Singh and Richard S Sutton. Reinforcement learning with replacing eligibility traces. *Machine learning*, 22(1-3):123–158, 1996.

[SSR97] Juan C Santamaría, Richard S Sutton, and Ashwin Ram. Experiments with reinforcement learning in problems with continuous state and action spaces. *Adaptive behavior*, 6(2):163–217, 1997.

[SSSN08] Qing Song, James C Spall, Yeng Chai Soh, and Jie Ni. Robust neural network tracking controller using simultaneous perturbation stochastic approximation. *Neural Networks, IEEE Transactions on*, 19(5):817–835, 2008.

[ST12] Warren L Stutzman and Gary A Thiele. *Antenna theory and design*. John Wiley & Sons, 2012.

[Ste97] Willi-Hans Steeb. *Matrix calculus and Kronecker product with applications and C++ programs*. World Scientific, 1997.

[Sto92] Anton Stoorvogel. *H-infinity control problem. A state space approach.* Prentice Hall., 1992.

[Str07] Julius Adams Stratton. *Electromagnetic theory*, volume 33. John Wiley & Sons, 2007.

[Sut84] Richard Stuart Sutton. Temporal credit assignment in reinforcement learning. 1984.

[Sut96] Richard S Sutton. Generalization in reinforcement learning: Successful examples using sparse coarse coding. *Advances in neural information processing systems*, pages 1038–1044, 1996.

[SW89] Sharad Singhal and Lance Wu. Training multilayer perceptrons with the extended Kalman algorithm. In *Advances in neural information processing systems*, pages 133–140, 1989.

[Swi96] Kevin Swingler. *Applying neural networks: a practical guide*. Morgan Kaufmann, 1996.

[SZ06] VE Semenov and NA Zharova. Thermal runaway and hot spots under controlled microwave heating. In *Advances in Microwave and Radio Frequency Processing*, pages 482–490. Springer, 2006.

[SZL+95] Jonas Sjöberg, Qinghua Zhang, Lennart Ljung, Albert Benveniste, Bernard Delyon, Pierre-Yves Glorennec, Håkan Hjalmarsson, and Anatoli Juditsky. Nonlinear black-box modeling in system identification: A unified overview. *Automatica*, 31(12):1691–1724, 1995.

[TBS10] Evangelos Theodorou, Jonas Buchli, and Stefan Schaal. Reinforcement learning of motor skills in high dimensions: A path integral approach. In *Robotics and Automation (ICRA), 2010 IEEE International Conference on*, pages 2397–2403. IEEE, 2010.

[TC99] ET Thostenson and T-W Chou. Microwave processing: Fundamentals and applications. *Com-*

posites Part A: Applied Science and Manufacturing, 30(9):1055–1071, 1999.

[tef] Introduction of teflon. http://www.engineering-toolbox.com/thermal-conductivity-d429.html. Accessed: 2015-02-06.

[Tes92] Gerald Tesauro. *Practical issues in temporal difference learning*. Springer, 1992.

[Tes95] Gerald Tesauro. Temporal difference learning and TD-Gammon. *Communications of the ACM*, 38(3):58–68, 1995.

[THN01] LMM Tijskens, MLATM Hertog, and Bart M Nicolaï. *Food process modelling*, volume 59. Woodhead Publishing, 2001.

[Tho97] Thomas L Thorpe. Vehicle traffic light control using SARSA. 1997.

[Tho98] Edward Lee Thorndike. Animal intelligence: An experimental study of the associate processes in animals. *American Psychologist*, 53(10):1125, 1998.

[Thr92] Sebastian B Thrun. Efficient exploration in reinforcement learning. 1992.

[Tia11] Tian Tian. Anisotropic thermal property measurement of carbon-fiber/epoxy composite materials. 2011.

[TK06] Jeffrey Travis and Jim Kring. *LabVIEW for Everyone: Graphical Programming Made Easy and Fun (National Instruments Virtual Instrumentation Series)*. Prentice Hall PTR, 2006.

[TMBO03] Claire J Tomlin, Ian Mitchell, Alexandre M Bayen, and Meeko Oishi. Computational techniques for the verification of hybrid systems. *Proceedings of the IEEE*, 91(7):986–1001, 2003.

[TP08] Peter Trebaticky̌ and Jiří Pospíchal. Neural network training with extended Kalman filter using graphics processing unit. In *Artificial Neural Networks-ICANN 2008*, pages 198–207. Springer, 2008.

[Tre05] Robert J Trew. High-frequency solid-state electronic devices. *Electron Devices, IEEE Transactions on*, 52(5):638–649, 2005.

[TTA09] Yoshimasa Tsuruoka, Jun'ichi Tsujii, and Sophia Ananiadou. Stochastic gradient descent training for L1-regularized log-linear models with cumulative penalty. In *Proceedings of the Joint Conference of the 47th Annual Meeting of the ACL and the 4th International Joint Conference on Natural Language Processing of the AFNLP: Volume 1-Volume 1*, pages 477–485. Association for Computational Linguistics, 2009.

[Tul07] Harry Tuller. Ionic conduction and applications. In *Springer Handbook of Electronic and Photonic Materials*, pages 213–228. Springer, 2007.

[TVR97] John N Tsitsiklis and Benjamin Van Roy. An analysis of temporal-difference learning with function approximation. *Automatic Control, IEEE Transactions on*, 42(5):674–690, 1997.

[vdB07] RAPM van den Bleek. MIMO H-infinity control design for the AVIS. 2007.

[VPAKL09] Draguna Vrabie, O Pastravanu, Murad Abu-Khalaf, and Frank L Lewis. Adaptive optimal control for continuous-time linear systems based on policy iteration. *Automatica*, 45(2):477–484, 2009.

[Wan09] Liuping Wang. *Model predictive control system design and implementation using MATLAB®*. springer, 2009.

[WCJ+12] Birron Mathew Weedy, Brian John Cory, N Jenkins, JB Ekanayake, and G Strbac. *Electric power systems*. John Wiley & Sons, 2012.

[WD92] Christopher JCH Watkins and Peter Dayan. Q-learning. *Machine learning*, 8(3-4):279–292, 1992.

[web] Introduction of waveguides. http://www.microwaveengservices.com/waveguide.html. Accessed: 2014-12-01.

[Wer92] Paul J Werbos. Approximate dynamic programming for real-time control and neural modeling. *Handbook of intelligent control: Neural, fuzzy, and adaptive approaches*, 15:493–525, 1992.

[WF05] Ian H Witten and Eibe Frank. *Data Mining: Practical machine learning tools and techniques*. Morgan Kaufmann, 2005.

[Whi94] Darrell Whitley. A genetic algorithm tutorial. *Statistics and computing*, 4(2):65–85, 1994.

[Wil09] John K Williams. Reinforcement learning of optimal controls. In *Artificial intelligence methods in the environmental sciences*, pages 297–327. Springer, 2009.

[Win99] RHS Winterton. Newton's law of cooling. *Contemporary Physics*, 40(3):205–212, 1999.

[WKK+14] A Wieckowski, P Korpas, M Krysicki, F Dughiero, M Bullo, F Bressan, and Christian Fager. Efficiency optimization for phase controlled multi-source microwave oven. *International Journal of Applied Electromagnetics and Mechanics*, 44(2):235–241, 2014.

[WMS92] Paul J Werbos, Thomas McAvoy, and Ted Su. Neural networks, system identification, and control in the chemical process industries. *Handbook of Intelligent ControlNeural, Fuzzy, and Adaptive approaches*, pages 283–356, 1992.

[WP97] Eric Walter and Luc Pronzato. Identification of parametric models. *Communications and Control Engineering*, 1997.

[wr3] Introduction of WR-340 waveguide. http://luxtem.com/wizhome/menu125.html. Accessed: 2015-03-10.

[WRK+12] Oliver Wiedenmann, Robert Ramakrishnan, E Kilic, Patrick Saal, Uwe Siart, and Thomas F Eibert. A multi-physics model for microwave heating in metal casting applications embedding a mode stirrer. In *Microwave Conference (GeMiC), 2012 The 7th German*, pages 1–4. IEEE, 2012.

[WS98] Marco Wiering and Jürgen Schmidhuber. Speeding up Q (λ)-learning. In *Machine Learning: ECML-98*, pages 352–363. Springer, 1998.

[WSF94] DA White, DA Sofge, and LA Feldkamp. Handbook of intelligent control: neural, fuzzy, and adaptive approaches. *IEEE Transactions on Neural Networks*, 5(5):852–852, 1994.

[WVDM00] Eric A Wan and Rudolph Van Der Merwe. The unscented Kalman filter for nonlinear estimation. In *Adaptive Systems for Signal Processing, Communications, and Control Symposium 2000. AS-SPCC. The IEEE 2000*, pages 153–158. IEEE, 2000.

[WWP00] Markus Weber, Max Welling, and Pietro Perona. *Unsupervised learning of models for recognition*. Springer, 2000.

[WYT12] Wei Wei, Hong-Ming Yin, and Juming Tang. An optimal control problem for microwave heating. *Nonlinear Analysis: Theory, Methods & Applications*, 75(4):2024–2036, 2012.

[XC02] SY Xu and TW Chen. Robust H-infinity control for uncertain stochastic systems with state delay. *IEEE Transactions on Automatic Control*, 47(12):2089–2094, 2002.

[XCH06] Qiang Xiong, Wen-Jian Cai, and Ming He. A practical decentralized PID auto-tuning method for tito systems under closed–loop control. *International Journal of Innovative Computing, Information and Control*, 2(2):305–322, 2006.

[Y+66] Kane S Yee et al. Numerical solution of initial boundary value problems involving Maxwells equations in isotropic media. *IEEE Trans. Antennas Propag*, 14(3):302–307, 1966.

[YF94] Ronald R Yager and Dimitar P Filev. Essentials of fuzzy modeling and control. *New York*, 1994.

[YG95] H.W. Yang and S. Gunasekaran. *Foundations and industrial applications of microwave and radio frequency fields: physical and chemical processes*, volume 492. John Wiley and Sons, 1995.

[YG04] H.W. Yang and S. Gunasekaran. Comparison of temperature distribution in model food cylinders based on Maxwell's equations and Lambert's law during pulsed microwave heating. *Journal of Food Engineering*, 64(4):445 – 453, 2004.

[ZL03] Yuanjin Zheng and Zhiping Lin. Recursive adaptive algorithms for fast and rapidly time-varying systems. *Circuits and Systems II: Analog and Digital Signal Processing, IEEE Transactions on*, 50(9):602–614, 2003.

[ZM03] Changjiu Zhou and Qingchun Meng. Dynamic balance of a biped robot using fuzzy reinforcement learning agents. *Fuzzy sets and systems*, 134(1):169–187, 2003.

[ZQY14] Ziyuan Zhang, Xiuchen Qiao, and Jianguo Yu. Microwave selective heating-enhanced reaction rates for mullite preparation from kaolinite. *RSC Advances*, 4(6):2640–2647, 2014.

[ZSWM00] Zheng Zhang, Scott Schwartz, Lukas Wagner, and Webb Miller. A greedy algorithm for aligning DNA sequences. *Journal of Computational biology*, 7(1-2):203–214, 2000.

A. Appendix

A.1. Derivation of Extended Kalman Filter

When the input vector $\mathbf{V}(k)$is not zero, the state parameter $A^n(k-1)$ is assumed to be constant like $A^n(k-1) = A^n_{\mathrm{ct}}$, where A^n_{ct} is the last updated value of A^n. Then the n-th MISO systems can be written as such as

$$\begin{aligned} Y^n(k) &= A^n_{\mathrm{ct}} \cdot Y^n(k-1) + \Psi^n(k-1), \\ \Psi^n(k-1) &= \mathbf{V}^{\mathrm{T}}(k-1)\,[\mathbf{\Phi}^n(k-1)]\,\mathbf{V}(k-1). \end{aligned} \tag{A.1}$$

At any time k, the value of $\Psi^n(k-1)$ can be calculated by

$$\Psi^n(k-1) = Y^n(k) - A^n_{\mathrm{ct}} \cdot Y^n(k-1), \tag{A.2}$$

and then this value is used to estimate $[\mathbf{\Phi}^n(k-1)]$ in equations A.1.

In order to make the derivation process more intuitive to understand, the original variables $\mathbf{V}(k-1)$, $\Psi^n(k-1)$ and $[\mathbf{\Phi}^n(k-1)]$ are replaced by new notations as

$$\mathbf{\Gamma}(k) = \mathbf{V}(k-1), \quad \Delta Y^n(k) = \Psi^n(k-1), \quad [\mathbf{\Lambda}^n(k)] = [\mathbf{\Phi}^n(k-1)]\,. \tag{A.3}$$

Then the second equation in A.1 can be rewritten as

$$\Delta Y^n(k) = \mathbf{\Gamma}^{\mathrm{T}}(k)\,[\mathbf{\Lambda}^n(k)]\,\mathbf{\Gamma}(k). \tag{A.4}$$

As in the linear RKF (equation 3.59), in EKF it is also assumed that the system fulfills the following equations

$$\begin{aligned} \Delta Y^n_{\mathrm{r}}(k) &= \mathbf{\Gamma}^{\mathrm{T}}(k)\,[\mathbf{\Lambda}^n(k)]\,\mathbf{\Gamma}(k) + \varsigma(k), \\ [\mathbf{\Lambda}^n(k)] &= [\mathbf{\Lambda}^n(k-1)] + [\varepsilon(k-1)]\,, \end{aligned} \tag{A.5}$$

where $\Delta Y_{\mathrm{r}}^{n}(k)$ is the real measured/calculated value of $\Delta Y^{n}(k)$, $\varsigma(k-1)$ is a zero-mean measurement/calculation error with the covariance σ^2 and $[\varepsilon(k-1)]$ is a zero-mean white noise with the covariance matrix $[\mathbf{\Omega}]$. The definitions of parameters in EKF is shown by table A.1.

Definition	Interpretation
$[\mathbf{\Lambda}_{\mathrm{e}}^{n}(k)] = E\big[[\mathbf{\Lambda}^{n}(k)] \,\big|\, \Delta Y_{\mathrm{r}}^{n}(k)\big]$	Estimation of $[\mathbf{\Lambda}^{n}(k)]$ based on the current measured temperature $\Delta Y_{\mathrm{r}}^{n}(k)$
$\big[\mathbf{\Lambda}_{\mathrm{p}}^{n}(k)\big] = E\big[[\mathbf{\Lambda}^{n}(k)] \,\big|\, \Delta Y_{\mathrm{r}}^{n}(k-1)\big]$	Prediction of $[\mathbf{\Lambda}^{n}(k)]$ based on the former measured temperature $\Delta Y_{\mathrm{r}}^{n}(k-1)$
$[\mathbf{e}_{\mathrm{e}}^{n}(k)] = [\mathbf{\Lambda}^{n}(k)] - [\mathbf{\Lambda}_{\mathrm{e}}^{n}(k)]$	Estimation error between $[\mathbf{\Lambda}^{n}(k)]$ and $[\mathbf{\Lambda}_{\mathrm{e}}^{n}(k)]$
$\big[\mathbf{e}_{\mathrm{p}}^{n}(k)\big] = [\mathbf{\Lambda}^{n}(k)] - \big[\mathbf{\Lambda}_{\mathrm{p}}^{n}(k)\big]$	Prediction error between $[\mathbf{\Lambda}^{n}(k)]$ and $\big[\mathbf{\Lambda}_{\mathrm{p}}^{n}(k)\big]$
$[\mathbf{P}_{\mathrm{e}}^{n}(k)] = E\left[[\mathbf{e}_{\mathrm{e}}^{n}(k)]\,[\mathbf{e}_{\mathrm{e}}^{n}(k)^{\mathrm{T}}]\right]$	Covariance matrix of the estimation error $[\mathbf{e}_{\mathrm{e}}^{n}(k)]$
$\big[\mathbf{P}_{\mathrm{p}}^{n}(k)\big] = E\left[\big[\mathbf{e}_{\mathrm{p}}^{n}(k)\big]\,\big[\mathbf{e}_{\mathrm{p}}^{n}(k)^{\mathrm{T}}\big]\right]$	Covariance matrix of the prediction error $\big[\mathbf{e}_{\mathrm{p}}^{n}(k)\big]$

Table A.1. Parameters and vectors used in EKF

According to the definitions in table A.1, the relationship between the estimation and the prediction of $\mathbf{\Lambda}^{n}(k)$ can be expressed by

$$
\begin{aligned}
\big[\mathbf{\Lambda}_{\mathrm{p}}^{n}(k)\big] &= E\big[[\mathbf{\Lambda}^{n}(k)] \,\big|\, \Delta Y_{\mathrm{r}}^{n}(k-1)\big] \\
&= E\big[[\mathbf{\Lambda}^{n}(k-1)] + [\varepsilon(k-1)] \,\big|\, \Delta Y_{\mathrm{r}}^{n}(k-1)\big] \\
&= E\big[[\mathbf{\Lambda}^{n}(k-1)] \,\big|\, \Delta Y_{\mathrm{r}}^{n}(k-1)\big] \\
&= [\mathbf{\Lambda}_{\mathrm{e}}^{n}(k-1)]
\end{aligned}
\tag{A.6}
$$

Then the prediction error matrix $\left[\mathbf{e}_{\mathrm{p}}^{n}(k)\right]$ can be rewritten as

$$\begin{aligned}\left[\mathbf{e}_{\mathrm{p}}^{n}(k)\right] &= \left[\mathbf{\Lambda}^{n}(k)\right] - \left[\mathbf{\Lambda}_{\mathrm{p}}^{n}(k)\right] \\ &= \left[\mathbf{\Lambda}^{n}(k)\right] - \left[\mathbf{\Lambda}_{\mathrm{e}}^{n}(k-1)\right] \\ &= \left[\mathbf{\Lambda}^{n}(k-1)\right] + \left[\varepsilon(k-1)\right] - \left[\mathbf{\Lambda}_{\mathrm{e}}^{n}(k-1)\right] \\ &= \left[\mathbf{\Lambda}^{n}(k-1)\right] - \left[\mathbf{\Lambda}_{\mathrm{e}}^{n}(k-1)\right] + \left[\varepsilon(k-1)\right] \\ &= \left[\mathbf{e}_{\mathrm{e}}^{n}(k-1)\right] + \left[\varepsilon(k-1)\right]\end{aligned} \tag{A.7}$$

Based on this result, the covariance matrix $\left[\mathbf{P}_{\mathrm{p}}^{n}(k)\right]$ can also be transferred into another form, such as

$$\begin{aligned}\left[\mathbf{P}_{\mathrm{p}}^{n}(k)\right] &= E\left[\left[\mathbf{e}_{\mathrm{p}}^{n}(k)\right]\left[\mathbf{e}_{\mathrm{p}}^{n}(k)\right]^{\mathrm{T}}\right] \\ &= E\left[\left(\left[\mathbf{e}_{\mathrm{e}}^{n}(k-1)\right] + \left[\varepsilon(k-1)\right]\right)\left(\left[\mathbf{e}_{\mathrm{e}}^{n}(k-1)\right] + \left[\varepsilon(k-1)\right]\right)^{\mathrm{T}}\right] \\ &= E\left[\left[\mathbf{e}_{\mathrm{e}}^{n}(k-1)\right]\left[\mathbf{e}_{\mathrm{e}}^{n}(k-1)\right]^{\mathrm{T}} + \left[\varepsilon(k-1)\right]\left[\varepsilon(k-1)\right]^{\mathrm{T}}\right] \\ &= \left[\mathbf{P}_{\mathrm{e}}^{n}(k-1)\right] + \left[\mathbf{\Omega}\right]\end{aligned} \tag{A.8}$$

As represented by equation 3.61, the update equation of $\left[\mathbf{\Lambda}_{\mathrm{e}}^{n}(k)\right]$ in the recursive system identification is represented by

$$\left[\mathbf{\Lambda}_{\mathrm{e}}^{n}(k)\right] = \left[\mathbf{\Lambda}_{\mathrm{p}}^{n}(k)\right] + \left[\mathbf{K}^{n}(k)\right]\left[\Delta Y_{\mathrm{r}}^{n}(k) - \mathbf{\Gamma}^{\mathrm{T}}(k)\left[\mathbf{\Lambda}_{\mathrm{p}}^{n}(k)\right]\mathbf{\Gamma}(k)\right] \tag{A.9}$$

Substituting the above expression into the definition of $\left[\mathbf{e}_{\mathrm{e}}^{n}(k)\right]$, the estimation error matrix can be written as

$$\begin{aligned}\left[\mathbf{e}_{\mathrm{e}}^{n}(k)\right] &= \left[\mathbf{\Lambda}^{n}(k)\right] - \left[\mathbf{\Lambda}_{\mathrm{e}}^{n}(k)\right] \\ &= \left[\mathbf{\Lambda}^{n}(k-1)\right] + \left[\varepsilon(k-1)\right] - \left[\mathbf{\Lambda}_{\mathrm{e}}^{n}(k)\right] \\ &= \left[\mathbf{\Lambda}^{n}(k-1)\right] + \left[\varepsilon(k-1)\right] - \left[\mathbf{\Lambda}_{\mathrm{p}}^{n}(k)\right] \\ &\qquad - \left[\mathbf{K}^{n}(k)\right]\left[\Delta Y_{\mathrm{r}}^{n}(k) - \mathbf{\Gamma}^{\mathrm{T}}(k)\left[\mathbf{\Lambda}_{\mathrm{p}}^{n}(k)\right]\mathbf{\Gamma}(k)\right] \\ &= \left[\mathbf{\Lambda}^{n}(k-1)\right] + \left[\varepsilon(k-1)\right] - \left[\mathbf{\Lambda}_{\mathrm{e}}^{n}(k-1)\right] \\ &\qquad - \left[\mathbf{K}^{n}(k)\right]\left[\Delta Y_{\mathrm{r}}^{n}(k) - \mathbf{\Gamma}^{\mathrm{T}}(k)\left[\mathbf{\Lambda}_{\mathrm{p}}^{n}(k)\right]\mathbf{\Gamma}(k)\right]\end{aligned} \tag{A.10}$$

The real measured/calculated $\Delta Y_{\mathrm{r}}^{n}(k)$ can be approximated by its Taylor series at $[\boldsymbol{\Lambda}^{n}(k)] = [\boldsymbol{\Lambda}_{\mathrm{p}}^{n}(k)]$, such as

$$\begin{aligned}
\Delta Y_{\mathrm{r}}^{n}(k) &= \boldsymbol{\Gamma}^{\mathrm{T}}(k)\left[\boldsymbol{\Lambda}_{\mathrm{p}}^{n}(k)\right]\boldsymbol{\Gamma}(k) + [\mathbf{J}^{n}(k)]\left[[\boldsymbol{\Lambda}^{n}(k)] - \left[\boldsymbol{\Lambda}_{\mathrm{p}}^{n}(k)\right]\right] + \varsigma(k) \\
&\quad + \mathrm{H.O.T.} \\
&\approx \boldsymbol{\Gamma}^{\mathrm{T}}(k)\left[\boldsymbol{\Lambda}_{\mathrm{p}}^{n}(k)\right]\boldsymbol{\Gamma}(k) + [\mathbf{J}^{n}(k)]\left[[\boldsymbol{\Lambda}^{n}(k)] - \left[\boldsymbol{\Lambda}_{\mathrm{p}}^{n}(k)\right]\right] + \varsigma(k) \\
&\approx \boldsymbol{\Gamma}^{\mathrm{T}}(k)\left[\boldsymbol{\Lambda}_{\mathrm{p}}^{n}(k)\right]\boldsymbol{\Gamma}(k) + [\mathbf{J}^{n}(k)]\left[\mathbf{e}_{\mathrm{p}}^{n}(k)\right] + \varsigma(k) \\
&\approx \boldsymbol{\Gamma}^{\mathrm{T}}(k)\left[\boldsymbol{\Lambda}_{\mathrm{p}}^{n}(k)\right]\boldsymbol{\Gamma}(k) + [\mathbf{J}^{n}(k)]\left[[\mathbf{e}_{\mathrm{e}}^{n}(k-1)] + [\boldsymbol{\varepsilon}(k-1)]\right] \\
&\quad + \varsigma(k),
\end{aligned} \tag{A.11}$$

where H.O.T. stands for higher order terms and $[\mathbf{J}^{n}(k)]$ is the Jacobian matrix defined by

$$[\mathbf{J}^{n}(k)] = \left.\frac{\partial \Delta Y_{\mathrm{r}}^{n}}{\partial [\boldsymbol{\Lambda}^{n}]}\right|_{[\boldsymbol{\Lambda}^{n}(k)] = [\boldsymbol{\Lambda}_{\mathrm{p}}^{n}(k)]} = \boldsymbol{\Gamma}(k)\boldsymbol{\Gamma}^{\mathrm{T}}(k). \tag{A.12}$$

Substituting the above expression A.11 into equation A.10, the estimation error matrix is further transferred into

$$\begin{aligned}
[\mathbf{e}_{\mathrm{e}}^{n}(k)] &= [\boldsymbol{\Lambda}^{n}(k-1)] + [\boldsymbol{\varepsilon}(k-1)] - [\boldsymbol{\Lambda}_{\mathrm{e}}^{n}(k-1)] \\
&\qquad - [\mathbf{K}^{n}(k)]\left[\Delta Y_{\mathrm{r}}^{n}(k) - \boldsymbol{\Gamma}^{\mathrm{T}}(k)\left[\boldsymbol{\Lambda}_{\mathrm{p}}^{n}(k)\right]\boldsymbol{\Gamma}(k)\right] \\
&= [\boldsymbol{\Lambda}^{n}(k-1)] + [\boldsymbol{\varepsilon}(k-1)] - [\boldsymbol{\Lambda}_{\mathrm{e}}^{n}(k-1)] - \varsigma(k)\,[\mathbf{K}^{n}(k)] \\
&\qquad - [\mathbf{K}^{n}(k)]\,[\mathbf{J}^{n}(k)]\left[[\mathbf{e}_{\mathrm{e}}^{n}(k-1)] + [\boldsymbol{\varepsilon}(k-1)]\right] \\
&= [\mathbf{e}_{\mathrm{e}}^{n}(k-1)] + [\boldsymbol{\varepsilon}(k-1)] - \varsigma(k)\,[\mathbf{K}^{n}(k)] \\
&\qquad - [\mathbf{K}^{n}(k)]\,[\mathbf{J}^{n}(k)]\left[[\mathbf{e}_{\mathrm{e}}^{n}(k-1)] + [\boldsymbol{\varepsilon}(k-1)]\right] \\
&= \left[[\mathbf{I}_{M}] - [\mathbf{K}^{n}(k)]\,[\mathbf{J}^{n}(k)]\right]\left[[\mathbf{e}_{\mathrm{e}}^{n}(k-1)] + [\boldsymbol{\varepsilon}(k-1)]\right] \\
&\qquad - \varsigma(k)\,[\mathbf{K}^{n}(k)] \\
&= \left[[\mathbf{I}_{M}] - [\mathbf{K}^{n}(k)]\,[\mathbf{J}^{n}(k)]\right]\left[\mathbf{e}_{\mathrm{p}}^{n}(k)\right] - \varsigma(k)\,[\mathbf{K}^{n}(k)],
\end{aligned} \tag{A.13}$$

where $[\mathbf{I}_{M}]$ is the identity matrix with the dimension $M \times M$.

Then finally the covariance matrix of the estimation error can be expressed by

$$
\begin{aligned}
[\mathbf{P}_{\mathrm{e}}^{n}(k)] &= E\left[[\mathbf{e}_{\mathrm{e}}^{n}(k)]\,[\mathbf{e}_{\mathrm{e}}^{n}(k)]^{\mathrm{T}}\right] \\
&= E\Bigg[\Big(\big([\mathbf{I}] - [\mathbf{K}^{n}(k)]\,[\mathbf{J}^{n}(k)]\big)\,\big[\mathbf{e}_{\mathrm{p}}^{n}(k)\big] - \varsigma(k)\,[\mathbf{K}^{n}(k)]\Big) \\
&\qquad \Big(\big([\mathbf{I}] - [\mathbf{K}^{n}(k)]\,[\mathbf{J}^{n}(k)]\big)\,\big[\mathbf{e}_{\mathrm{p}}^{n}(k)\big] - \varsigma(k)\,[\mathbf{K}^{n}(k)]\Big)^{\mathrm{T}}\Bigg] \\
&= \big([\mathbf{I}] - [\mathbf{K}^{n}(k)]\,[\mathbf{J}^{n}(k)]\big)\,\big[\mathbf{P}_{\mathrm{p}}^{n}(k)\big]\,\big([\mathbf{I}] - [\mathbf{K}^{n}(k)]\,[\mathbf{J}^{n}(k)]\big)^{\mathrm{T}} \\
&\qquad + \sigma^{2}\,[\mathbf{K}^{n}(k)]\,[\mathbf{K}^{n}(k)]^{\mathrm{T}} \\
&= \big[\mathbf{P}_{\mathrm{p}}^{n}(k)\big] - [\mathbf{K}^{n}(k)]\,[\mathbf{J}^{n}(k)]\,\big[\mathbf{P}_{\mathrm{p}}^{n}(k)\big] \\
&\qquad - \big[\mathbf{P}_{\mathrm{p}}^{n}(k)\big]\,[\mathbf{J}^{n}(k)]^{\mathrm{T}}\,[\mathbf{K}^{n}(k)]^{\mathrm{T}} \\
&\qquad + [\mathbf{K}^{n}(k)]\,[\mathbf{J}]^{n}(k)\,\big[\mathbf{P}_{\mathrm{p}}^{n}(k)\big]\,[\mathbf{J}^{n}(k)]^{\mathrm{T}}\,[\mathbf{K}^{n}(k)]^{\mathrm{T}} \\
&\qquad + \sigma^{2}\,[\mathbf{K}^{n}(k)]\,[\mathbf{K}^{n}(k)]^{\mathrm{T}}
\end{aligned}
\tag{A.14}
$$

The objective of EKF is to minimize the estimation error $[\mathbf{e}_{\mathrm{e}}^{n}(k)]$ regarding the gain matrix $[\mathbf{K}^{n}(k)]$, which in other words, is equivalent to find an appropriate matrix $[\mathbf{K}^{n}(k)]$ minimizing the covariance matrix $[\mathbf{P}_{\mathrm{e}}^{n}(k)]$.

Following the principle of

$$
[\mathbf{K}^{n}(k)] = \arg_{[\mathbf{K}]} \min\,[\mathbf{P}_{\mathrm{e}}^{n}(k)] \;\Rightarrow\; \frac{\partial\,[\mathbf{P}_{\mathrm{e}}^{n}(k)]}{\partial\,[\mathbf{K}^{n}(k)]} = 0, \tag{A.15}
$$

the gain matrix is obtained as

$$
[\mathbf{K}^{n}(k)] = \big[\mathbf{P}_{\mathrm{p}}^{n}(k)\big]\,[\mathbf{J}^{n}(k)]^{\mathrm{T}}\left([\mathbf{J}^{n}(k)]\,\big[\mathbf{P}_{\mathrm{p}}^{n}(k)\big]\,[\mathbf{J}^{n}(k)]^{\mathrm{T}} + \sigma^{2}\,[\mathbf{I}_{M}]\right)^{-1}. \tag{A.16}
$$

Using the above expression, equation A.14 can be further simplified as

$$\begin{aligned}
[\mathbf{P}_{\mathrm{e}}^{n}(k)] &= \big([\mathbf{I}_M]-[\mathbf{K}^n(k)]\,[\mathbf{J}^n(k)]\big)\,\big[\mathbf{P}_{\mathrm{p}}^{n}(k)\big]-\Big(\big[\mathbf{P}_{\mathrm{p}}^{n}(k)\big]\,[\mathbf{J}^n(k)]^{\mathrm{T}} \\
&\qquad -[\mathbf{K}^n(k)]\,[\mathbf{J}^n(k)]\,\big[\mathbf{P}_{\mathrm{p}}^{n}(k)\big]\,[\mathbf{J}^n(k)]^{\mathrm{T}} \\
&\qquad -\sigma^2\,[\mathbf{K}^n(k)]\Big)\,[\mathbf{K}^n(k)]^{\mathrm{T}} \\
&= \big([\mathbf{I}_M]-[\mathbf{K}^n(k)]\,[\mathbf{J}^n(k)]\big)\,\big[\mathbf{P}_{\mathrm{p}}^{n}(k)\big]-\Big[\big[\mathbf{P}_{\mathrm{p}}^{n}(k)\big]\,[\mathbf{J}^n(k)]^{\mathrm{T}} \\
&\qquad -[\mathbf{K}^n(k)]\,\big([\mathbf{J}^n(k)]\,\big[\mathbf{P}_{\mathrm{p}}^{n}(k)\big]\,[\mathbf{J}^n(k)]^{\mathrm{T}}+\sigma^2\big)\Big]\,[\mathbf{K}^n(k)]^{\mathrm{T}} \\
&= \big([\mathbf{I}_M]-[\mathbf{K}^n(k)]\,[\mathbf{J}^n(k)]\big)\,\big[\mathbf{P}_{\mathrm{p}}^{n}(k)\big]-\Big(\big[\mathbf{P}_{\mathrm{p}}^{n}(k)\big]\,[\mathbf{J}^n(k)]^{\mathrm{T}} \\
&\qquad -\big[\mathbf{P}_{\mathrm{p}}^{n}(k)\big]\,[\mathbf{J}^n(k)]^{\mathrm{T}}\Big)\,[\mathbf{K}^n(k)]^{\mathrm{T}} \\
&= \big([\mathbf{I}_M]-[\mathbf{K}^n(k)]\,[\mathbf{J}^n(k)]\big)\,\big[\mathbf{P}_{\mathrm{p}}^{n}(k)\big]
\end{aligned} \tag{A.17}$$

Combining equations A.6, A.8, A.12, A.14 and A.16, the entire update rule in EKF is represented as in the following.

Prediction part:

$$\begin{aligned}
\big[\mathbf{\Lambda}_{\mathrm{p}}^{n}(k)\big] &= [\mathbf{\Lambda}_{\mathrm{e}}^{n}(k-1)]\,, \\
\big[\mathbf{P}_{\mathrm{p}}^{n}(k)\big] &= [\mathbf{P}_{\mathrm{e}}^{n}(k-1)]+[\mathbf{\Omega}]\,.
\end{aligned}$$

Estimation part:

$$\begin{aligned}
[\mathbf{K}^n(k)] &= \big[\mathbf{P}_{\mathrm{p}}^{n}(k)\big]\,[\mathbf{J}^n(k)]^{\mathrm{T}}\Big([\mathbf{J}^n(k)]\,\big[\mathbf{P}_{\mathrm{p}}^{n}(k)\big]\,[\mathbf{J}^n(k)]^{\mathrm{T}}+\sigma^2\,[\mathbf{I}]\Big)^{-1}, \\
[\mathbf{\Lambda}_{\mathrm{e}}^{n}(k)] &= \big[\mathbf{\Lambda}_{\mathrm{p}}^{n}(k)\big]+[\mathbf{K}^n(k)]\,\Big(\Delta Y_{\mathrm{r}}^{n}(k)-\mathbf{\Gamma}^{\mathrm{T}}(k)\,\big[\mathbf{\Lambda}_{\mathrm{p}}^{n}(k)\big]\,\mathbf{\Gamma}(k)\Big), \\
[\mathbf{P}_{\mathrm{e}}^{n}(k)] &= \big([\mathbf{I}_M]-[\mathbf{K}^n(k)]\,[\mathbf{J}^n(k)]\big)\,\big[\mathbf{P}_{\mathrm{p}}^{n}(k)\big]\,,
\end{aligned}$$

with

$$[\mathbf{J}^n(k)] = \mathbf{\Gamma}(k)\mathbf{\Gamma}^{\mathrm{T}}(k).$$

Replacing the new defined variables by the original variables, the co-variance matrices of the estimation and the prediction errors are

$$\begin{aligned}
&[\mathbf{P}_{\mathrm{e}}^{n}(k-1)] \\
&\quad = E\left[\left([\mathbf{\Phi}^{n}(k-1)] - [\mathbf{\Phi}_{\mathrm{e}}^{n}(k-1)]\right)\left([\mathbf{\Phi}^{n}(k-1)] - [\mathbf{\Phi}_{\mathrm{e}}^{n}(k-1)]\right)^{\mathrm{T}}\right] \\
&\left[\mathbf{P}_{\mathrm{p}}^{n}(k-1)\right] \\
&\quad = E\left[\left([\mathbf{\Phi}^{n}(k-1)] - \left[\mathbf{\Phi}_{\mathrm{p}}^{n}(k-1)\right]\right)\left([\mathbf{\Phi}^{n}(k-1)] - \left[\mathbf{\Phi}_{\mathrm{p}}^{n}(k-1)\right]\right)^{\mathrm{T}}\right].
\end{aligned} \tag{A.18}$$

Then at each time k, the effective heating matrix $[\mathbf{\Phi}^{n}]$ can be updated as following.

Prediction part:

$$\begin{aligned}
\left[\mathbf{\Phi}_{\mathrm{p}}^{n}(k-1)\right] &= [\mathbf{\Phi}_{\mathrm{e}}^{n}(k-2)], \\
\left[\mathbf{P}_{\mathrm{p}}^{n}(k-1)\right] &= [\mathbf{P}_{\mathrm{e}}^{n}(k-2)] + [\mathbf{\Omega}].
\end{aligned}$$

Estimation part:

$$\begin{aligned}
[\mathbf{K}^{n}(k-1)] &= \left[\mathbf{P}_{\mathrm{p}}^{n}(k-1)\right][\mathbf{J}^{n}(k-1)]^{\mathrm{T}}\left([\mathbf{J}^{n}(k-1)]\left[\mathbf{P}_{\mathrm{p}}^{n}(k-1)\right]\right. \\
&\qquad \left.[\mathbf{J}^{n}(k-1)]^{\mathrm{T}} + \sigma^{2}[\mathbf{I}_{M}]\right)^{-1}, \\
[\mathbf{\Phi}_{\mathrm{e}}^{n}(k-1)] &= \left[\mathbf{\Phi}_{\mathrm{p}}^{n}(k-1)\right] + [\mathbf{K}^{n}(k-1)]\left(\Psi_{\mathrm{r}}^{n}(k-1)\right. \\
&\qquad \left. - \mathbf{V}^{\mathrm{T}}(k-1)\left[\mathbf{\Phi}_{\mathrm{p}}^{n}(k-1)\right]\mathbf{V}(k-1)\right), \\
[\mathbf{P}_{\mathrm{e}}^{n}(k-1)] &= \left([\mathbf{I}_{M}] - [\mathbf{K}^{n}(k-1)][\mathbf{J}^{n}(k-1)]\right)\left[\mathbf{P}_{\mathrm{p}}^{n}(k-1)\right]
\end{aligned}$$

with

$$[\mathbf{J}^{n}(k-1)] = \mathbf{V}(k-1)\mathbf{V}^{\mathrm{T}}(k-1).$$

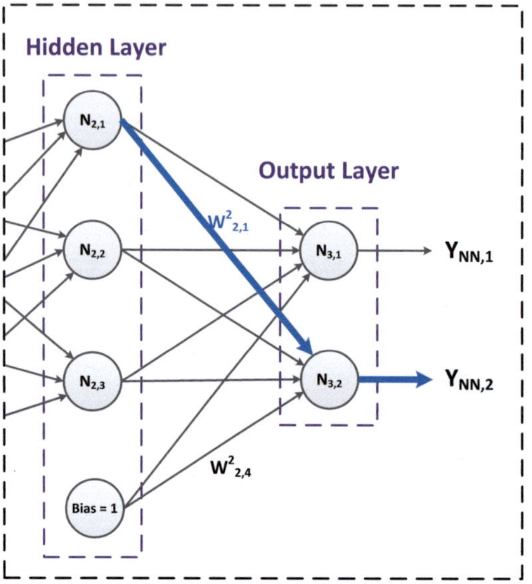

Figure B.1. Error flow in the output layer.

A.2. Derivation of Backpropagation Algorithm

Case 1: For weights connecting to the output layer

If the destination of the link is in the output layer, such as the blue link in figure B.1, any error of the weight only affects one output and the corresponding weight update is simple and straightforward. For the destination node j in the output layer, according to the definition in table 3.3 its input can be represented by

$$z_j^{\mathrm{o}}(k) = \sum_{i=1}^{N_L} w_{j,i}^{\mathrm{o}}(k) x_i^L(k), \tag{B.19}$$

where $x_i^L(k)$ is the input from the i-th node of the L-th hidden layer. Then the output of this node is

$$Y_{\mathrm{NN},j}(k) = f_{\mathrm{o}}(z_j^{\mathrm{o}}(k)), \tag{B.20}$$

where f_{o} represents the activation function of the output layer.

The corresponding update rule of any weight $w_{j,i}^{\mathrm{o}}(k)$ connecting with this node is

$$w_{j,i}^{\mathrm{o}}(k+1) = w_{j,i}^{\mathrm{o}}(k) + \Delta w_{j,i}^{\mathrm{o}}(k),$$

where

$$\begin{aligned} \Delta w_{j,i}^{\mathrm{o}}(k) &= -\eta \frac{\partial J(k)}{\partial w_{j,i}^{\mathrm{o}}(k)} \\ &= -\eta \frac{\partial J(k)}{\partial Y_{\mathrm{NN},j}(k)} \frac{\partial Y_{\mathrm{NN},j}(k)}{\partial f_{\mathrm{o}}(z_j^{\mathrm{o}}(k))} \frac{\partial f_{\mathrm{o}}(z_j^{\mathrm{o}}(k))}{\partial z_j^{\mathrm{o}}(k)} \frac{\partial z_j^{\mathrm{o}}(k)}{\partial w_{j,i}^{\mathrm{o}}(k)}. \end{aligned} \tag{B.21}$$

Substituting following expressions into the above equation (for each node in this layer)

$$\begin{aligned} \frac{\partial J(k)}{\partial Y_{\mathrm{NN},j}(k)} &= Y_{\mathrm{NN},j}(k) - Y_{\mathrm{d},j}(k), \\ \frac{\partial Y_{\mathrm{NN},j}(k)}{\partial f_{\mathrm{o}}(z_j^{\mathrm{o}}(k))} &= 1, \\ \frac{\partial f_{\mathrm{o}}(z_j^{\mathrm{o}}(k))}{\partial z_j^{\mathrm{o}}(k)} &= f_{\mathrm{o}}'(z_j^{\mathrm{o}}(k)), \\ \frac{\partial z_j^{\mathrm{o}}(k)}{\partial w_{j,i}^{\mathrm{o}}(k)} &= x_i^L(k), \end{aligned}$$

then equation B.21 will be transformed into

$$\begin{aligned} \Delta w_{j,i}^{\mathrm{o}}(k) &= -\eta \cdot \big(Y_{\mathrm{NN},j}(k) - Y_{\mathrm{d},j}(k)\big)\, f_{\mathrm{o}}'(z_j^{\mathrm{o}}(k))\, x_i^L(k) \\ &= -\eta \cdot \delta_j^{\mathrm{o}}(k)\, x_i^L(k), \\ \delta_j^{\mathrm{o}}(k) &= \big(Y_{\mathrm{NN},j}(k) - Y_{\mathrm{d},j}(k)\big) \cdot f_{\mathrm{o}}'(z_j^{\mathrm{o}}(k)), \end{aligned} \tag{B.22}$$

where $\delta_j^{\mathrm{o}}(k)$ is the local gradient equivalent to $\partial J(k)/\partial z_j^{\mathrm{o}}(k)$. When the activation function in the output layer is the identity function as

$f_o(z_j^o(k)) = z_j^o(k)$, the completed update equation is

$$\begin{aligned} w_{j,i}^o(k+1) &= w_{j,i}^o(k) - \eta \cdot \delta_j^o(k)\, x_i^L(k), \\ &= w_{j,i}^o(k) - \eta \cdot \delta_j^o(k)\, x_i^L(k). \end{aligned} \tag{B.23}$$

Case 2: For weights only connecting to hidden layers

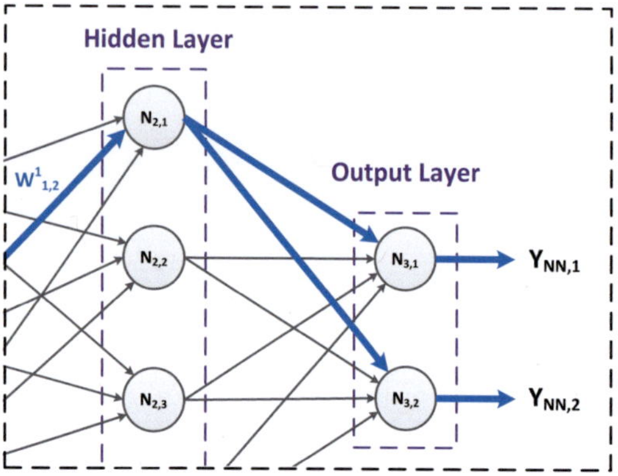

Figure B.2. Error flows from hidden layers to the output layer.

When both ends of the link are connected to nodes in hidden layers, the update of its weight becomes complicated. There is no specified desired output for nodes in hidden layers to directly calculate the error $\left(Y_{\mathrm{NN},j}(k) - Y_{\mathrm{d},j}(k)\right)$. In addition, all nodes in the output layer have to be considered because the error flows from the node in the hidden layer to all nodes in the output layer, such as indicated by the blue arrows in figure B.2.

Using a similar equation as in the first case to calculate $\Delta w_{j,i}^o(k)$, for the L-th layer there is

$$\begin{aligned} \Delta w_{j,i}^L(k) &= -\eta \frac{\partial J(k)}{\partial w_{j,i}^L(k)} = -\eta \frac{\partial J(k)}{\partial z_j^L(k)} \frac{\partial z_j^L(k)}{\partial w_{j,i}^L(k)} \\ &= -\eta \cdot \delta_j^L(k)\, x_i^{L-1}(k). \end{aligned} \tag{B.24}$$

By applying the chain rule, the local gradient $\delta_j^L(k)$ can be calculated as

$$\delta_j^L(k) = \sum_{p=1}^{N} \frac{\partial J(k)}{\partial f_\mathrm{o}(z_p^\mathrm{o}(k))} \frac{\partial f_\mathrm{o}(z_p^\mathrm{o}(k))}{\partial z_p^\mathrm{o}(k)} \frac{\partial z_p^\mathrm{o}(k)}{\partial g(z_j^L(k))} \frac{\partial g(z_j^L(k))}{\partial z_j^L(k)}. \tag{B.25}$$

where N is the number of nodes in the output layer.

According to the notations shown in figure 3.9 and the definitions in B.22, the terms in the above equation B.25 are expressed as

$$\begin{aligned} \frac{\partial J(k)}{\partial f_\mathrm{o}(z_p^\mathrm{o}(k))} &= \delta_p^\mathrm{o}(k), \\ \frac{\partial f_\mathrm{o}(z_p^\mathrm{o}(k))}{\partial z_p^\mathrm{o}(k)} &= 1, \\ \frac{\partial z_p^\mathrm{o}(k)}{\partial g(z_j^L(k))} &= w_{p,j}^\mathrm{o}(k), \\ \frac{\partial g(z_j^L(k))}{\partial z_j^L(k)} &= g'(z_j^L(k)). \end{aligned} \tag{B.26}$$

Substituting these expressions into the above equation B.25, it is rewritten as

$$\begin{aligned} \delta_j^L(k) &= \sum_{p=1}^{N} \delta_p^\mathrm{o}(k)\, w_{p,j}^\mathrm{o}(k)\, g'(z_j^L(k)) \\ &= g'(z_j^L(k)) \sum_{p=1}^{N} \delta_p^\mathrm{o}(k)\, w_{p,j}^\mathrm{o}(k). \end{aligned} \tag{B.27}$$

Equations B.24 and B.27 can be extended into a more general case where the node j is in former hidden layers [Hay98], such as

$$\begin{aligned} \Delta w_{j,i}^l(k) &= -\eta \cdot \delta_j^l(k)\, x_i^l(k) \\ &= -\eta \cdot \left[g'(z_j^{l+1}(k)) \sum_{p=1}^{N_{l+1}} \delta_p^{l+1}(k)\, w_{p,j}^{l+1}(k), \right] x_i^l(k). \end{aligned} \tag{B.28}$$

KARLSRUHER FORSCHUNGSBERICHTE AUS DEM
INSTITUT FÜR HOCHLEISTUNGSIMPULS- UND MIKROWELLENTECHNIK
(ISSN 2192-2764)

Die Bände sind unter www.ksp.kit.edu als PDF frei verfügbar oder als Druckausgabe bestellbar.

Band 1 **MATTHIAS BERINGER**
Design Studies towards a 4 MW 170 GHz Coaxial-Cavity Gyrotron. 2011
ISBN 978-3-86644-663-2

Band 2 **JENS FLAMM**
Diffraction and Scattering in Launchers of
Quasi-Optical Mode Converters for Gyrotrons. 2012
ISBN 978-3-86644-822-3

Band 3 **MATTIA DEL GIACCO**
Investigation of Fretting Wear of Cladding Materials in Liquid Lead. 2013
ISBN 978-3-86644-960-2

Band 4 **AMITAVO ROY CHOUDHURY**
Investigations of After Cavity Interaction in Gyrotrons
Including the Effect of Non-uniform Magnetic Field. 2013
ISBN 978-3-7315-0129-9

Band 5 **MICHAEL BETZ**
The CERN Resonant WISP Search (CROWS). 2014
ISBN 978-3-7315-0199-2

Band 6 **ANDREAS SCHLAICH**
Time-dependent spectrum analysis of high power gyrotrons. 2015
ISBN 978-3-7315-0375-0

Band 7 **DHIDIK PRASTIYANTO**
Temperature- and Time-Dependent Dielectric Measurements
and Modelling on Curing of Polymer Composites. 2016
ISBN 978-3-7315-0424-5

Band 8 **YIMING SUN**
Adaptive and Intelligent Temperature Control of Microwave
Heating Systems with Multiple Sources. 2016
ISBN 978-3-7315-0467-2